?

Tornado Alley

Tornado Alley

New York   Oxford   1999

OXFORD UNIVERSITY PRESS

Monster Storms of the Great Plains

HOWARD B. BLUESTEIN

Oxford University Press

Oxford   New York
Athens   Auckland   Bangkok   Bogotá   Buenos Aires
Calcutta   Cape Town   Chennai   Dar es Salaam   Delhi
Florence   Hong Kong   Istanbul   Karachi   Kuala Lumpur
Madrid   Melbourne   Mexico City   Mumbai   Nairobi   Paris
São Paulo   Singapore   Taipei   Tokyo   Toronto   Warsaw

and associated companies in
Berlin   Ibadan

Copyright © 1999 by Oxford University Press

Published by Oxford University Press, Inc.,
198 Madison Avenue, New York, New York 10016

Oxford is a registered trademark of Oxford University Press

Library of Congress Cataloging-in-Publication Data
Bluestein, Howard B.
Tornado alley : monster storms of the Great Plains / by Howard B. Bluestein.
p. cm.   Includes bibliographical references and index.
ISBN 0-19-510552-4
1. Tornadoes.   2. Tornadoes—Pictorial works.
3. Storms.   4. Storms—Pictorial works.   I. Title.
QC955.B58   1998   551.55'3—DC21   97-48265

1 3 5 7 9 8 6 4 2

Printed in Hong Kong
on acid-free paper

Contents

Preface

THE STIMULUS FOR WRITING this book can probably be traced back to my contact with Wendall Mordy, a cloud physicist, back in the early 1970s. I recall viewing his magnificent collection of cloud slides at his home in Coral Gables, Florida, where I was subsequently inspired to begin my own collection of cloud photographs. I had experimented with aerial cloud photography as early as the late 1950s, when I was in grade school. When I was an undergraduate majoring in electrical engineering at the Massachusetts Institute of Technology (MIT) during the late 1960s, I took a photography course in Doc Edgerton's laboratory and did quite a bit of work in black and white, in which I both chronicled the turbulent times of the late sixties and experimented with cloud photography. It was not until I was a graduate student in meteorology at MIT in the early 1970s that I began to accumulate my own collection of color slides. The collection began in earnest while I was working during the summer at the National Hurricane Research Laboratory (which is now the Hurricane Research Division of the Atlantic Oceanographic and Meteorological Laboratory). I continued to add to my collection while participating in numerous field experiments. I am a firm believer that in order to study a meteorological phenomenon properly, you must actually experience it and appreciate it aesthetically.

The photographs that I took, up to and including the 1981 storm season, were taken with a Miranda Sensomat 35-mm single-lens reflex camera. From that time until now I have been using one of two Nikon FM 35-mm single-lens reflex cameras. Most of my photographs were shot using a 50-mm f1.8 lens, some with a 28-mm f2.8 wide-angle lens, and a few with telephoto lenses having focal lengths as long as 200 mm. My film of choice since 1970 has been Kodachrome 25 (formerly Kodachrome II). In the 1970s and early 1980s I occasionally used Kodachrome 64. I prefer to use a slow, fine-grain film; under poor lighting conditions I shoot my pictures with my camera mounted on a tripod. The contrast of some daylight photographs of clouds was enhanced with a polarizing filter. All of the photographs in this book are mine unless otherwise noted.

Although I wrote the drafts of the manuscript for this book during a portion of a sabbatical leave from the University of Oklahoma when I was at the Mesoscale and Microscale Meteorology Division at the National Center for Atmospheric Research in Boulder, Colorado, the book is based in part on an invited presentation entitled "A Review of Storm Intercept Observations: Early 1970s to Now," which I delivered at the American Meteorological Society's seventeenth Conference on Severe Local Storms in St. Louis on October 5, 1993; it is also based in part on an article entitled "Riders on the Storm: Stalking Tornadoes to Learn How They Are Born," which I wrote for the April 1995 issue of *The Sciences*. Some of the topics discussed in this book have also appeared in an interview with Jeff Rosenfield, a former managing editor of *Weatherwise*, in the April/May 1996 issue of the magazine ("Spin Doctor: Talking Tornadoes with Howard Bluestein").

The purpose of this book is to describe, both photographically and scientifically, tornadoes and related severe-storm phenomena and to present a historical account of research done in the last half century aimed at discovering the characteristics of tornadoes and why they form. The text is written for the educated layperson with some knowledge of science. Physical explanations are given in lieu of detailed equations to avoid boring the nonmathematically inclined. The nonscientific layperson is advised, however, to skip most of the physical explanations also, and to focus on the historical narrative and the photographs. The professional scientist or potential student of meteorology, on the other hand, is encouraged to seek more detailed information, including the relevant equations, in the list of references given at the end of the book. Amateur severe-weather enthusiasts and professionals are encouraged to browse the sites on the World Wide Web listed in Appendix C to learn more about the weather and current activities going in the severe-thunderstorm arena. Current weather information and computer forecasts are available at some of the Web sites; other sites have information on current activities related to severe-storm research.

I opted to take a historical approach in discussing the topic of tornadoes, severe thunderstorms, and related phenomena in order to avoid a dry, textbook rendition of simply laying out all the "facts" as we currently know them. When certain weather phenomena and research tools come up naturally in the narrative, I then digress to explain them in more detail. My philosophy is similar to that I took when teaching a graduate-level course on mesoscale meteorology in the fall of 1991 to science operation officers from the National Weather Service at the Cooperative Program for Operational Meteorology, Education, and Training (COMET) in Boulder, Colorado. In the COMET course, we conducted case studies on workstations and stopped every once in a while to lecture on the dynamics of the features the students observed in analyses of data displayed on the workstations. I have included in the history of severe-storm and tornado research many anecdotal flashbacks. The flashbacks, which in my opinion are necessary, may appear confusing to some readers. The paths taken by meteorologists studying severe weather have not always been linear; many branches have been investigated simultaneously, and in some instances there have been paths revisited with new technology.

It is clear from a look at papers on severe-storm research published over the last forty to fifty years that technology has played a major role in advancing our knowledge. The development of radar technology and digital computer systems has been largely responsible for driving our research efforts. The emerging technologies mentioned near the end of the book will undoubtedly be the primary impetus for severe-storm research in the next five to ten years. It is likely that in the future, as in the past, breakthroughs in technology, serendipitous observations, and the hard work of some key individuals will lead to the significant advances in our field.

The reader should be cautioned that my awareness of severe-storm research began in the mid-1970s and that the historical perspective given of early projects is gleaned mainly from journal articles, a few books, and discussions and recollections of discussions with people who actually experienced the early years. It is as if I had not been alive before the mid-1970s: I was not conscious of what had gone on before my severe-storm innocence was lost. My treatment of history after the mid-1970s is based upon actual experience and is therefore my own personal perspective from my position at the University of Oklahoma in Norman, Oklahoma. I trust that others might tell the story of severe-storm research somewhat differently.

Of foremost importance, however, is the awe that one feels in the presence of the violent yet magnificent severe-storm phenomena that Nature reveals to us, regardless of individual perspectives. Even if someday we come to understand very precisely how and why tornadic storms form and behave, there will always be the thrill of experiencing them. The backdrops of the phenomena—in the expanses of the Great Plains, in the tropical splendor of south Florida and its adjacent waters, and in the alpine beauty of the Rocky Mountains—are as breathtaking as the phenomena themselves. I hope that the story and the photographs in this book stimulate the reader as much as seeing the real thing stimulated me.

Shoo-bop, shoo-bop
Howie "Cb" Bluestein
Boulder, Colorado
August 1997

Acknowledgments

I WOULD NOT have been able to write this book and collect severe-storm-related photographs without the long-term financial support I received from research grants from the National Science Foundation in Washington, D.C., and the National Severe Storms Laboratory (NSSL) in Norman, Oklahoma. The University of Oklahoma in Norman, the Mesoscale and Microscale Meteorology Division at the National Center for Atmospheric Research (NCAR) in Boulder, Colorado, and the Hurricane Research Division (formerly the National Hurricane Research Laboratory) of the Atlantic Oceanographic and Meteorological Laboratory in Miami all provided some measure of support, both intellectually and financially. I owe particular thanks to Frederick Sanders, professor emeritus at MIT, for starting me out on my career and to Edwin Kessler, former director of NSSL, for stimulating me to get involved in severe-storm research; I also thank Lance Bosart at the State University of New York at Albany, Robert Burpee at the National Hurricane Center in Miami, and Joseph Golden in Washington, D.C., for their continued friendship and support, and Jeff Kimpel at NSSL for getting me started storm chasing. My colleagues and friends Richard Rotunno and Morris Weisman from NCAR have been instrumental in my journey through the land of tornado and severe thunderstorm research and provided useful comments and stimulating discussions on several topics detailed in this book. The National Geographic Society provided support for the Waterspout Expedition in 1993. The National Aeronautics and Space Administration provided support for a lidar experiment in 1981. My many other colleagues and friends, especially those from the Norman, Boulder, and Miami communities, also contributed in some way to this book. The student crews who participated enthusiastically during the storm-intercept projects over the years and the personnel at NSSL and NCAR who made operations possible are gratefully acknowledged. Particular thanks are in order to the following: Herbert Stein, who for the last six years has kept our storm-intercept vehicle safely on the road; Donald Burgess, who over the years has been a stimulus for using Doppler-radar data to study severe storms; Al Bedard, who designed TOTO; Gregory Byrd and Eugene "Bill" McCaul, who made the portable radiosonde project work; Wesley Unruh, who collaborated for many years and gave many free hours of his time to work on the portable Doppler-radar project; John Umbarger of the Los Alamos National Laboratory who supported the work of Wes and his group; and Robert McIntosh and his group at the University of Massachusetts, who collaborated on the millimeter-wavelength mobile-radar project. Kelvin Droegemeier, Don MacGorman, and Josh Wurman provided comments on a few sections of the book relevant to their research efforts. Leslie Forehand at NCAR assisted with some of the literature search. Some of the figures were drafted by the NCAR Graphics group and by the Instructional Technology Service group at the University of Oklahoma. Victoria Kaharl and Joyce Berry spent many hours editing the manuscript and making helpful suggestions to make this book understandable to the layperson. I finally thank my mother and father, whose nurturing has ultimately allowed me to pursue a career in which I can enjoy myself almost all the time, weather permitting, and my wife, Kathleen Welch, who has gracefully put up with the eccentricities of a weather nut and his postmodern tornado rhetoric.

1

The Frontier Overhead

*Nihil est in intellectu quod
non antea fuerit in sensu.*
(All knowledge of the world
must rest finally on one's
sensory experience.)
—*John Locke*

I WAS A YOUNG CHILD playing
outside our house near Boston under a hazy yellow June sky in 1953 when
my mother summoned me inside because a tornado had been reported in
Worcester, about forty miles to the west. As further inducement to getting
me in the house, she told me that tornadoes snatch children up into the air
and abduct them. We didn't get many tornadoes in Massachusetts, but I
knew what happened to Dorothy and her dog, Toto, in *The Wizard of Oz*. As
it turned out, my mother needn't have worried. But in Worcester, the tornado killed ninety people, injured twelve hundred, and caused some $52
million (in 1953 dollars) in property damage.

That year was a big one in the United States for tornadoes: A total of 516
people were killed and many hundreds more were injured. Most of the car-
nage took place in three days from tornadoes in Waco, Texas (May 11), Flint,
Michigan (June 8), and Worcester (June 9).

In those days, the U.S. government's Weather Bureau did not issue tornado forecasts very often, at least not publicly. My mother knew there was
a tornado nearby because she heard about it on television.

Tornado forecasting can be traced back at least to the mid-1880s, when
the meteorologist J.P. Finley of the U.S. Army Signal Service (later called

I

the Signal Corps) dared to suggest that tornadoes were predictable. In 1883, however, the U.S. government banned the word *tornado* from forecasts to avoid panicking the masses. As the chief Signal Service officer explained in his report for 1887: "It is believed that the harm done by such a prediction would be greater than that which results from the tornado itself."

The ban was lifted in 1886, only to be reinstated the following year. The ban was lifted again in 1938, but only occasionally was the word *tornado* used and then only in forecasts issued to officials involved in disaster-relief efforts.

On March 20, 1948, a tornado struck Tinker Air Force Base, near Oklahoma City, and inflicted some $10 million in property losses. Five days later, when similar atmospheric conditions again presented themselves, two meteorologists at Tinker, Major Ernest Fawbush and Captain Robert Miller, issued a tornado warning, although this was not made public. A tornado did indeed strike, causing about $6 million in damages.

It was a lucky forecast. The odds of two tornadoes striking essentially the same place less than a week apart are infinitesimal. Perhaps it was this rare coincidence, which could not have escaped the meteorologists, as well as their incredible luck that sparked an unusual two-way exchange of government weather experts. During the following two years, the U.S. Weather Bureau in Kansas City invited Fawbush and Miller to discuss their forecasting procedures, and Tinker Air Force Base invited research meteorologists from Washington, D.C., to Missouri for talks. Of course, Fawbush and Miller had no special forecasting technique other than to examine weather maps and to determine that the maps looked the way they did five days before, when there was a tornado. The ultimate consensus of these meetings was that tornadoes really could not be forecast with much accuracy and therefore there was no good reason to change the existing policy of keeping tornado forecasts under wraps.

It wasn't until March 1952 that the U.S. Weather Bureau (the old Signal Corps) began to issue tornado forecasts publicly. Rather than cause panic, the warnings drew a good deal of criticism for their inaccuracy and lack of precision.

Our ability to predict tornadoes is still rudimentary. At best, we can say that in eight to twelve hours there might be some storms that might produce one or more tornadoes over a very broad area, sometimes covering several states. An hour or so in advance we can narrow down the area that might be affected, but we still cannot say whether a storm will bring a tornado.

Much about tornadoes remains mysterious. Tornadoes are often called one of the last frontiers of atmospheric science. We know so little about them because they are so hard to observe and study. Tornadoes are elusive. Most people have never seen one. My experience—more than twenty years of studying and chasing tornadoes—attests to the fact that the chances of catching one are remarkably slim.

Will Keller, a Kansas farmer, says he peered into the heart of a tornado's funnel cloud. According to his account, first published in the journal

Monthly Weather Review in 1930, the tornado appeared in the late afternoon of June 22, 1928. Keller says:

I was out in my field with my family looking over the ruins of our wheat crop which had just been completely destroyed by a hailstorm. I noticed an umbrella-shaped cloud in the west and southwest, and from its appearance, suspected that there was a tornado in it. The air had that peculiar oppressiveness which nearly always precedes the coming of a tornado.

But my attention being on other matters, I did not watch the approach of the cloud. However, its nearness soon caused me to take another look at it. I saw at once that my suspicions were correct, for hanging from the greenish-black base of the cloud was not just one tornado but three.

One of the tornadoes was already perilously near and apparently headed directly for our place. I lost no time therefore in hurrying with my family to our cyclone cellar.

The family had entered the cellar and I was in the doorway just about to enter and close the door when I decided that I would take a last look at the approaching tornado. I have seen a number of these things and have never become panic-stricken when near them. So I did not lose my head now, though the approaching tornado was indeed an impressive sight.

The surrounding country is level and there was nothing to obstruct the view. There was little or no rain falling from the cloud. Two of the tornadoes were some distance away and looked to me like great ropes dangling from the clouds, but the near one was shaped more like a funnel with ragged clouds surrounding it. It appeared to be much larger and more energetic than the others and it occupied the central position of the cloud, the great cumulus dome being directly over it.

As I paused to look, I saw that the lower end, which had been sweeping the ground, was beginning to rise. I knew what that meant, so I kept my position. I knew that I was comparatively safe and I knew that if the tornado again dipped, I could drop down and close the door before any harm could be done.

Steadily the tornado came on, the end gradually rising above the ground. I could have stood there only a few seconds but so impressed was I with what was going on that it seemed a long time. At last the great shaggy end of the funnel hung directly overhead. Everything was as still as death. There was a strong gassy odor and it seemed that I could not breathe. There was a screaming, hissing sound coming directly from the end of the funnel. I looked up and to my astonishment, I saw right up into the heart of the tornado. There was a circular opening in the center of the funnel, about 50 or 100 feet in diameter, and extending straight upward for a distance of at least one half mile, as best I could judge under the circumstances. The walls of this opening were of rotating clouds and the hole was made brilliantly visible by constant flashes of lightning which zigzagged from side to side. . . .

Around the lower rim of the great vortex small tornadoes were constantly forming and breaking away. These looked like tails as they writhed their way around the end of the funnel. It was these that made the hissing noise.

I noticed that the direction of rotation of the great whirl was anticlockwise, but the small twisters rotated both ways—some one way and some another.

The opening was entirely hollow except for something which I could not exactly make out, but suppose that it was a detached wind cloud. This thing was in the center and was moving up and down.

The tornado was not travelling at a great speed. I had plenty of time to get a good view of the whole thing, inside and out. . . . Its course was not in a straight line but zigzagged across the country in a general northeasterly direction.

After it passed my place, it again dipped and struck and demolished the house and barn of a farmer by the name of Evans. The Evans family, like ourselves, had been out looking over their hailed-out wheat and saw the tornado coming. Not having time to reach their cellar, they took refuge under a small bluff that faced to the leeward of the approaching tornado. They lay down flat on the ground and caught hold of some plum bushes which fortunately grew within their reach. As it was, they felt themselves lifted from the ground. Mr. Evans said that he could see the wreckage of his house, among it being the cook stove going round and round over his head. The eldest child, a girl of 17, being the most exposed, had her clothing completely torn off, but none of the family were hurt.

I am not the first one to lay claims to having seen the inside of a tornado. I remember that in 1915, a tornado passed near Mullinville and a hired man on a farm over which the tornado passed had taken refuge in the barn. As the tornado passed over the barn, the door was blown open and the man saw up into it, and this one, like the one I saw, was hollow and lit up by lightning. As the hired man was not well known no one paid much attention to what he said.

Keller seems to be a credible observer and, at least to a meteorologist who studies tornadoes, a very lucky one. The only time I've ever seen inside a tornado funnel was on a radar display. Nor have I ever heard any "screaming" or "hissing" from the hundred or more tornadoes I've witnessed. I do know that you can't tell simply from looking at a thunderstorm whether it will spawn a tornado. And Keller was overly confident; another tornado a seemingly safe distance away may not have given him the chance to crawl into his cellar (or 'fraidy hole, as I've heard it called in the plains) and shut the door. Not all tornadoes are associated with "peculiar oppressiveness" or humidity, as Keller says. Nor is there necessarily any gassy odor associated with tornadoes. Perhaps what Keller smelled was gas from a line break.

—☙❧—

The Fujita-Pearson Tornado Intensity Scale, formulated in 1971, rates tornadoes from F1 to F6 (Table 1.1). This scale estimates tornado intensity based solely on the degree of damage caused by tornadoes. It was designed to connect smoothly with the Beaufort scale, devised in 1806 by Sir Francis

Table 1.1   The Fujita F scale

F number	F-scale damage specification
F0	18–32 m/sec (40–72 mph): Light damage; some damage to chimneys; break branches off trees; push over shallow-rooted trees; damage signboards
F1	33–49 m/sec (73–112 mph): Moderate damage; the lower limit (73 mph) is the beginning of hurricane wind speed; peel surface off roofs; mobile homes pushed off foundations or overturned; moving autos pushed off the road
F2	50–69 m/sec (113–157 mph): Considerable damage; roofs torn off frame houses; mobile homes demolished; boxcars pushed over; large trees snapped or uprooted; light-object missiles generated
F3	70–92 m/sec (158–206 mph): Severe damage; roofs and some walls torn off well-constructed houses; trains overturned; most trees in forest uprooted; heavy cars lifted off ground and thrown
F4	93–116 m/sec (207–260 mph): Devastating damage; well-constructed houses leveled; structures with weak foundations blown off some distance; cars thrown and large missiles generated
F5	117–142 m/sec (261–318 mph): Incredible damage; strong frame houses lifted off foundations and carried considerable distance to disintegrate; automobile-sized missiles fly through the air in excess of 100 meters; trees debarked; incredible phenomena will occur
F6–F12	143 m/sec to Mach 1, the speed of sound (319–700 mph): The maximum wind speeds of tornadoes are not expected to reach the F6 wind speeds*

*Brian Fiedler at the University of Oklahoma believes that vertical speeds in this range are possible in suction vortices.

Beaufort, a British naval officer. The Beaufort scale gives wind speed from 0, meaning calm, to 12, for hurricane-force winds. The Fujita scale picks up where the Beaufort scale ends and eventually reaches Mach 1, the speed of sound. An F6 tornado, according to Fujita, is "inconceivable." Tetsuya "Ted" Fujita is a meteorologist at the University of Chicago who would become one of the founding fathers of severe-storm studies. Fujita had begun his career in postwar Japan studying the pattern of damage from the atomic bomb attack on Hiroshima.

The Fujita scale is not perfect. Estimating tornado intensity based on damage alone, not actual wind measurements, is risky. The structural integrity of the things hit by the tornado is often not very well known. A well-built structure can withstand very high winds, while a poorly built structure can suffer devastating damage even from less intense winds. Most tornado damage is a result of pressure induced by the wind. The pressure exerted on an object is proportional to the square of the wind speed and to a factor that depends on the shape of the object and the wind direction. Two identical objects that feel the wind from different angles may suffer different amounts of damage. Furthermore, many tornadoes occur over open country, where there is nothing to damage.

Reports of tornado damage that imply incredibly high winds can be misleading. For example, people are amazed to learn that straw has been driven into wood posts and trees during a tornado. Experiments performed in the laboratory with a pneumatic gun have demonstrated that straw splinters can indeed be driven into soft wood by winds as weak as 60–70 mph. Much higher wind speeds are needed to drive the splinters into hard wood.

There have also been reports of tornadoes stripping fowl of their feathers. In 1842, an American professor named Elias Loomis, presumably a scientist, conducted an experiment with the objective of determining how high the wind speed must be to blow all the feathers off a chicken. He did this because "the stripping of fowls [during tornadoes]," Loomis said, "attracted much attention." He continued:

> In order to determine the velocity needed to strip these feathers . . . [a] six-pounder was loaded with five ounces of powder, and for a ball, a chicken just killed. The gun was pointed vertically upwards and fired. The feathers rose twenty or thirty feet, and were scattered by the wind. On examination, they were found to be pulled out clean, the skin seldom adhering to them. The body was torn into small fragments, only a part of which could be found. The velocity was 341 miles per hour. A fowl, then, forced through the air with this velocity is torn entirely to pieces; with a less [sic] velocity, it is probable most of the feathers might be pulled out without mutilating the body.

More recent experiments indicate that the force needed to remove feathers varies widely, depending on the chicken's condition.

There have been reports of tornadoes passing over ponds and subsequently "raining out" frogs. It is not difficult to imagine how a tornado can suck up the frogs along with some of the water and deposit them elsewhere, but I know of no experiments that have attempted to determine how strong a tornado must be to account for the frog-rain phenomenon.

During the summer of 1997 two people and the cottage in which they were sheltered were lifted in a tornado and deposited in a nearby lake. The pair managed to swim back to shore and were hospitalized. One was reported to have remarked, "There was a hell of a bang, and I can remember saying 'Where in the hell did all this water come from?'"

I've been as close as a quarter of a mile to a tornado, but I've never seen airborne cows. I have seen dead cows on the ground after a tornado has passed through, and I have seen houses that have been knocked off their foundations. It may be that cows, houses, and people do fly through the violently rotating winds of tornadoes, but we usually can't see them because of the dark cloud of dirt and debris that tornadoes usually whip up.

What would it take to lift a cow and toss it through the air? Consider that the speed at which air is rising in a tornado increases with height, from zero at the ground to some high value, say 100 mph, about three hundred feet above the ground. For a cow to become airborne, it would have to be lifted high enough so that the gravitational force acting downward on its body is more than counteracted by the upward drag force it would experience as it rises. But the upward wind speed is probably not great enough to lift the animal off the ground. More likely is that the horizontal wind speed at the top of the cow will be high enough to blow it laterally and flip it over; this might cause the cow to tumble and bounce high enough to reach the level at which it becomes truly airborne.

The recent movie *Twister*, in which there is a flying cow, is a comic-book version of what meteorologists do. We do not put ourselves inside tornadoes to take measurements. This is simply not possible. Nor, as a rule, are we suicidal in our quest to observe a tornado firsthand.

In terms of destructive power, tornadoes are the most violent storms on earth. Hurricanes and typhoons pack winds of 200 mph at most; the top speeds of tornadoes have been estimated at 300 mph. No instrument to measure wind speed directly has ever survived a strong tornado.

Storms that spawn tornadoes may also bring hail plummeting from aloft at 70 mph, destroying crops, shattering windows, and pummeling anything and anyone in their path. The biggest hailstone I know of was found in Coffeyville, Kansas, on September 3, 1970; it weighed 1.67 pounds and was 17.5 inches around.

Tornadoes occur throughout the world, over mountains, above plains and coastlines, in valleys, and over oceans, but they are by far most frequent in a hundred-mile-swath in the central United States that has come to be called "Tornado Alley." In this stretch, which extends from northern Texas and the Texas Panhandle through Oklahoma and Nebraska (Fig. 1.1), there

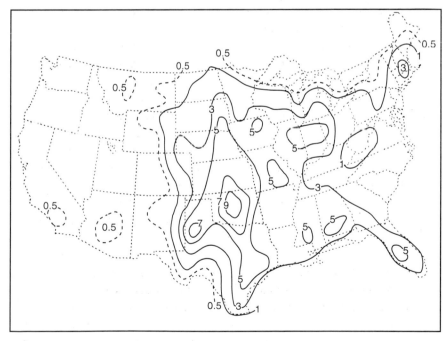

1.1 *Frequency of occurrence of tornadoes (1950–1976) in the United States as number per year within a circle of radius of one degree of latitude-longitude. A maximum of 10.5 is located in central Oklahoma. (Adapted from Schaefer et al. 1980 and Kelly et al. 1978; courtesy of the American Meteorological Society. Copyright 1980.) (A newer version that includes tornadoes up to and including 1989 is qualitatively similar.)*

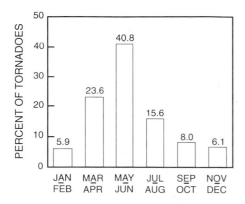

1.2 *Bimonthly climatology of tornado occurrence (percentage of all tornadoes) in the United States. (Adapted from Kelly et al. 1978; courtesy of the American Meteorological Society.) (A newer version that includes tornadoes up to and including 1989 is qualitatively similar.)*

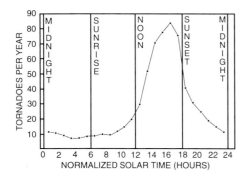

1.3 *Tornado occurrence as a function of time of day (1950–1976) in the United States. (Adapted from Kelly et al. 1978; courtesy of the American Meteorological Society.) (A newer version that includes tornadoes up to and including 1989 is qualitatively similar.)*

are at least five tornadoes every year within a circle of radius of one degree of latitude-longitude (i.e., in a circular area approximately 50 miles across). Statistics on their frequency, however, must be viewed with a certain amount of skepticism. The data are probably biased toward urban areas, since it is more likely that tornadoes in remote parts of the world go unreported. Furthermore, counting tornadoes is not as straightforward as it may seem. Sometimes a series of tornadoes accounts for one long, continuous damage path. Sometimes a tornado will spin up for only a matter of seconds, disappear, and then reappear. If we included every such disappearing act logged in our records, we could produce a much higher frequency rate for tornadoes.

There is no time of year and no time of day that tornadoes do not occur, but most are reported in the spring and early summer (Fig. 1.2), from midafternoon to early evening (Fig. 1.3), when the atmospheric temperatures at the ground are highest. However, tornadoes have occurred even over snow-covered ground. On Dec. 2, 1970, a tornado hit Timpanogos Divide, Utah, buried under 38 inches of snow.

Most tornadoes rotate cyclonically, which is counterclockwise in the Northern Hemisphere and clockwise in the Southern Hemisphere. Exceptions to this have been documented. Although Northern Hemisphere tornadoes usually move from the southwest to the northeast, a tornado can move in any direction. The same tornado can change direction in midstream; the axes of some tornadoes appear to be stationary. A tornado typically lasts for less than half an hour, sometimes only several minutes or even seconds; its damage path may be as long as a hundred miles and as wide as two miles.

Tornadoes affect very small portions of our planet and are short-lived, but they kill scores of people in the United States alone every year and cause hundreds of millions of dollars in property damage. (Statistics for other places are not as well known). Before 1950 the average annual death count from tornadoes in the United States often exceeded a hundred; in recent years, the figure has been well under that number, due probably to improved warnings and increased public awareness.

But there are always exceptions. In what has come to be called the Palm Sunday outbreak of April 11, 1965, a family of tornadoes, at least thirty-seven and perhaps more than fifty, killed 256 people in Iowa, Wisconsin, Illinois, Indiana, Michigan, and Ohio. In another outbreak, on April 3–4, 1974, 148 tornadoes killed three hundred people in thirteen states stretching from Alabama to Michigan. On June 9, 1984, tornadoes killed four hundred in the former Soviet Union. In Bangladesh, a tornado killed more than five hundred people and injured some thirty-three thousand on May 13, 1996.

Tornadoes come in so many shapes and sizes and guises that they seem to defy definition. What exactly is a tornado? Meteorologists are still trying to answer this question. The word *tornado* probably comes from the Spanish *tornar*, "to turn," and *tornado*, "thunderstorm." A tornado is commonly defined as a violently rotating column of air hanging from a tall, bubbly cloud from which rain is falling (a cumulonimbus). But I have seen torna-

1.4   *Horizontal tornado funnel on April 22, 1985, looking west from Weatherford, Oklahoma.*

does hanging from clouds that contain no rain. And "column" implies that the vortex of rotating air is vertically oriented; some tornadoes are stretched out horizontally on their sides (Fig. 1.4). Although most, but not all, tornadoes are associated with powerful thunderstorms, I've seen tornadoes come from weak thunderstorms. And not all tornadoes are spawned by thunderstorms. Hurricanes and typhoons, when they make landfall, can spawn tornadoes. To complicate the matter further, other meteorological phenomena—such as landspouts, waterspouts, and dust devils—are also violently rotating columns of air. The latter is not considered a tornado, but the first two are.

So, what is a tornado? It is a violently rotating column of air, which may not be oriented vertically, that comes from beneath the base of a thunderstorm or a rapidly growing towering cumulus cloud.

Tornado Research: Early History

One of the earliest studies of tornadoes was a small pilot program conducted in 1950 near Washington, D.C. The goal was to test the hypothesis of Morris Tepper of the U.S. Weather Bureau, who suggested that the intersection of two pressure-jump lines—features on a weather map indicating a rapid rise in barometric pressure—is a preferred zone of tornado development.

The experiment, dubbed the Tornado Project, was soon expanded to include parts of Kansas and Oklahoma. Apparently, tornadoes were frustratingly scarce, as the project's investigators noted. They wrote in their report: "Unfortunately for meteorological knowledge, the setting up of the Tornado Project system seems to have provided the people of Kansas with the best tornado insurance they ever had. For at the present writing (June), there have been no tornadoes in the 'arc' area during 1952."

It was finally determined that the majority of tornadoes are not related to intersecting pressure-jump lines. But the experiment marked an early use of mesonetworks, closely deployed surface instruments that measure such variables as temperature and barometric pressure. Surface measuring devices used operationally are spaced a hundred miles or more apart. In the mesonetwork, the instruments were one mile to ten miles apart, giving a more in-depth picture of atmospheric conditions.

In the late 1950s meteorologists studied movies taken of tornadoes, usually by nonscientists. The first scientific analysis was made, in 1957, by Walter Hoecker of the U.S. Weather Bureau. Hoecker had acquired a film of a tornado ripping through Dallas, Texas, on April 2, 1957. Frame by frame, the film faithfully tracked swirling chunks of debris. Because he knew the distance from the camera to the tornado, he could compute how far the debris moved from frame to frame. Knowing these distances and the time interval between frames, he could estimate the wind speed. Using this kind of elementary geometry, Hoecker estimated wind speeds as high as 170 mph. Nobody had ever estimated the wind speed of a tornado before, although there had been talk that tornadoes could be supersonic, with wind speeds in excess of 700 mph.

The same year, on June 20, a number of local citizens photographed a tornado in Fargo, North Dakota. One of the shots in this sequence of spectacular still photos showed an airborne automobile. The photographs made their way to Ted Fujita. From the Fargo pictures, he determined the relationship between a tornado and the architecture of cloud features associated with its parent storm. For his analysis, published in 1960, he coined the terms *wall cloud, tail cloud,* and others. But it would be years and many more photographs before the usefulness of his terminology would be appreciated. After all, the Fargo tornado was only one event; the scientific establishment was loath to apply Fujita's analysis to all tornadic storms, which are thunderstorms that spawn tornadoes.

In the 1950s and 1960s aerial photographs of tornado damage were analyzed. A photograph of damage to a cornfield in Nebraska in 1955 shows a pattern of loops of flattened stalks. Meteorologist E. L. Van Tassel attributed the loops to scratch marks from large debris. Fujita later hypothesized, in 1970, that similar cycloidal (looped) or scalloped patterns of damage in grain fields from the infamous 1965 Palm Sunday outbreak were caused by "suction spots" that rotated around the tornado (Fig. 1.5). These spots supposedly sucked up debris from the ground as they rotated around the center of the tornado. What suction spots do is still controversial; it is not clear whether debris is sucked up or simply blown over during a tornado.

On April 9, 1953, a weather radar in Champaign, Illinois, picked up a

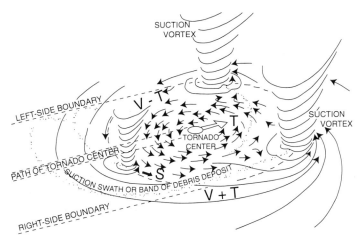

1.5 *Fujita's model of a tornado with multiple suction vortices. T, S, and V indicate the translational speed of the tornado, the translational speed of each suction vortex, and the rotational speed around the tornado core, respectively. (Adapted from Fujita 1981; courtesy of the American Meteorological Society. Copyright 1981.)*

hook-shaped appendage on the southwest side of a tornadic storm (Fig. 1.6). Such a hook, which had been seen before but not captured by a camera, was conjectured to be associated with the circulation of a tornado. Meteorologists would later learn that not all hooks are associated with tornadoes, nor are all tornadoes associated with hooks.

Meteorologists began using conventional (non-Doppler) radar in the late 1940s to study thunderstorm structure and behavior. It had been discovered during World War II that radar detects not only aircraft but also precipitation. Aircraft were small points of light on the radar screen and precipitation appeared as pulsating blotches. To the military, this was interference or "noise": It was not the enemy and did not shoot you, although precipitation in the form of large hail could indeed cause harm.

To the meteorologist, radar illuminates the invisible by rendering it visible; it shows us the raindrops and cloud droplets up to several hundred miles away. At radar's centimeter wavelengths (which correspond to frequencies on the order of 3 to 10 GHz), raindrops (and cloud droplets) act as Rayleigh scatterers. Rayleigh scattering occurs when the wavelength is long compared to the width of the scatterers; the energy of the backscattered radiation decreases rapidly with increased wavelength. Precipitation backscatters to the radar, and this is the echo. (Electromagnetic radiation from radar is also scattered to the front and sides, but meteorologists usually aren't concerned with this because the radiation does not return to the radar directly, or at all.) The largest raindrops backscatter the largest fraction of the energy, while the tiniest raindrops and the cloud droplets backscatter the least. The actual power received back at the radar device is given by the weighted sum of the diameters of the drops or droplets raised to the sixth power. In other words, a raindrop 4 mm in diameter backscatters sixty-four times more powerfully than a raindrop 2 mm in diameter.

Cloud droplets differ from raindrops in that the latter are large enough

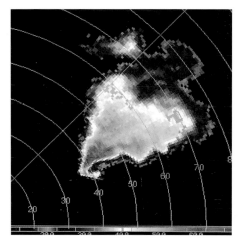

1.6 *Hook echo in a tornadic thunderstorm near Arcadia, Oklahoma, May 17, 1981, at 5:58 P.M. (time given here and elsewhere in local standard time or in daylight savings time, when appropriate). Radar reflectivity is plotted in dBZ. Range markers are at 10-km intervals from NSSL's Cimarron radar. A tornado formed a few minutes later. The shape of the yellow radar echo looks a bit like a scorpion or the tip of Cape Cod. (Courtesy of David Dowell, School of Meteorology, University of Oklahoma, and Don Burgess, formerly at NSSL.)*

to have a fall velocity relative to the air. Cloud droplets, much smaller, do not—they just blow along with the wind.

Radar bands in the 3–10 GHz frequency range are used because those with longer wavelengths are not as sensitive to raindrops and cloud droplets and require cumbersome antennas. Much more of the radiation from shorter wavelengths is absorbed by raindrops, so more of the beam is attenuated, and as a result, the sensitivity to rain at great distances is markedly diminished.

Conventional radar can detect only the intensity of precipitation in a volume of air. Doppler radar can detect this *and* the rate of motion—that is, the speed of the scatterers along the line of sight of the radar. Doppler radar makes use of the Doppler effect, as expounded by Christian Doppler, a nineteenth-century physicist. The frequency of waves is changed by their motion relative to that of an observer. For example, consider the sound waves of a train's whistle: The pitch of the whistle rises as the train approaches and becomes lower as the train recedes. Light waves are another example. The visible spectrum of light from approaching stars shifts into the blue region of the spectrum, while the visible light from receding stars shifts into the red region of the spectrum. This red shift has been associated with an expanding universe in which virtually all galaxies more than three million light-years away are receding. Consider, finally, electromagnetic waves when they backscatter from raindrops and cloud droplets. The frequency of backscattered radiation shifts up if the raindrops or cloud droplets move toward the radar but shifts down if the water moves away from the radar. Doppler radar can sense whether the raindrops are moving toward or away from the radar, and this tells us how fast the wind is blowing because it is the wind that propels the raindrops.

Suppose that raindrops, dust, and other debris are carried along by the wind in a tornado. The antenna of the Doppler radar concentrates electromagnetic energy into a narrow beam. If the beam is concentrated along an arc of 1 degree, then at a range of thirty miles, the beam will be spread out over a distance of about half a mile. Since all but the largest tornadoes are no wider than a mile or so, the radar beam will encompass most of the tornado. Since the left side of the vortex has scatterers (i.e., raindrops, dust, debris, etc.) that are approaching the radar, and the right side of the vortex has scatterers receding from the radar, the shifts in frequency of radiation backscattered to the radar are both up and down. In fact, if we simply compute the average Doppler shift, weighted by the intensity of the radiation at each frequency shift, we find that it is close to zero and the tornado would not be detected. But if we consider the spread of frequency shifts detected, the *Doppler velocity spectrum,* then we would see that some backscatterers are moving away and some are moving toward the radar. The maximum wind speeds in the direction of the line of sight from the radar can thus be determined (Fig. 1.7). Note that Doppler radar can sense only the motion along the line of sight of the radar, while the photogrammetric analysis technique can yield only the motion perpendicular to the line of sight.

The first Doppler-radar measurements of a tornado were made in 1958 by U.S. Weather Bureau meteorologists R. Smith and D. Holmes. They obtained a Doppler radar from the U.S. Navy, set it up in Wichita, Kansas, and then waited. Just over a year later, on June 10, the radar detected a tornado in El Dorado twenty-five miles away. What luck that a tornado happened to touch down within range of the first Doppler radar used for meteorological purposes! The likelihood of a tornado striking within thirty miles of a given location during a given spring is extremely small. Although at the time it was not possible to discriminate between approaching and receding velocities, the wind spectra obtained by this radar detected maximum velocities (that is, wind speeds in the line-of-sight direction) of 200 mph. Smith and Holmes's radar was a continuous-wave (CW) radar at 3-cm wavelength. CW radars send electromagnetic energy continuously. An aural analogy might be a coyote that howls continuously in a canyon; the echo of the howl mixes with the howl going out. Thus, one cannot determine the delay between the time energy is radiated and the time backscattered energy reaches the radar. If the coyote were to howl in bursts, then, given the speed of sound, the time delay between each burst and its echo could be used to determine the distance to the canyon wall.

A major drawback of these first Doppler-radar observations was that range information could not be obtained, owing to the way the signal was processed. Pulsed-Doppler radars can determine range. But in 1958 it was not yet possible to obtain both range and velocity from CW radar. The Doppler radar was not used after the first successful tornado measurements, and it was more than a decade before next major advances in Doppler-radar technology were put to meteorological use.

Besides the remote measurements made by radar and the photographs and movies, measurements of wind and pressure were made serendipitously, when a tornado happened to pass near a weather station. Meteorologist Ed Brooks reported in 1949 that tornadoes often appear within low-pressure areas that are about ten miles across. He named these gyres "tornado cyclones."

Only data from calibrated instruments should be trusted, and it is not always clear which of the early data are reliable. Anemometers tend to get blown away by tornadoes. One trace from an anemometer showed peak gusts of 145 mph when a tornado passed just north of a weather station during the Palm Sunday tornado outbreak. On May 27, 1896, a pressure deficit (deviation from ambient pressure) of 82 mb was measured very near the center of a tornado in St. Louis. A waterspout passing over a ship in 1958 dropped the pressure by 21 mb. On May 24, 1962, a pressure deficit of 34 mb was recorded near what was called a "tornado cyclone" in Newton, Kansas. An NSSL mesonet site also recorded a 10-mb pressure deficit some 1,300–2,600 feet from the center of a tornado on April 30, 1970. Although we don't know how good these measurements are, the largest credible pressure drop is probably about 100 mb, which is about 10 percent of the total atmospheric pressure near the ground. Pressure deficits in intense oceanic extratropical cyclones can be higher than 50 mb; pressure deficits in tropical cyclones can reach almost 100 mb. In the case of the extratropical and tropical cyclones, the large pressure drops cover much greater horizontal distances than those of tornadoes.

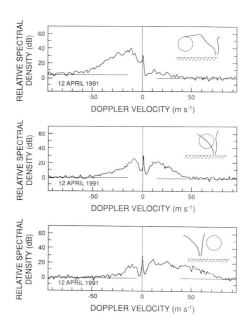

1.7 *Doppler wind spectra in a tornado on April 12, 1991, obtained with data from a portable Doppler radar looking (top) to the left, (middle) centered near, and (bottom) to the right of the tornado. Approaching (receding) velocities are negative (positive). Maximum Doppler velocities are indicated where the relative spectral density enters the noise floor from lower velocities. For example, in the bottom panel, receding Doppler velocities as high as 80 m/sec are indicated. (From Bluestein et al. 1993; courtesy of the American Meteorological Society. Copyright 1993.)*

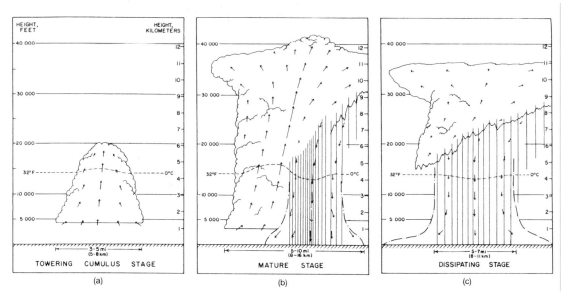

1.8 *The Byers-and-Braham idealized life cycle of a thunderstorm cell: (a) the cumulus stage, (b) the mature stage, and (c) the dissipating stage. Arrows indicate the sense of air motion. Vertical lines indicate precipitation. (Based upon a figure from C. Doswell. From Bluestein 1993.)*

Thunderstorm Research: Early History

Our understanding of tornadoes also benefited from other early work that focused on thunderstorms, the usual parents of tornadoes. Meteorologist Horace Byers and graduate student Roscoe Braham from the University of Chicago conducted a detailed study of thunderstorms near Orlando, Florida, in the summer of 1946, and near Wilmington, Ohio, the following summer. Their study, called the Thunderstorm Project, was prompted by a number of thunderstorm-related airplane crashes. Their objective was to try to learn what to expect when flying in and around thunderstorms.

During the course of the Thunderstorm Project, which would become the prototype for modern meteorological field experiments, techniques were developed to make *in situ* measurements. It also may have been the first time pilots flew systematically into thunderstorms and let their aircraft ride out the storms by allowing the storms to take control. Besides giving the pilots a wild and even dangerous ride, this allowed them to estimate the intensity of updrafts and downdrafts. The project also used pulsed conventional radar, surface instruments, instrumented weather balloons (rawinsondes), and instruments such as thermometers on the aircraft themselves.

The main results of the Thunderstorm Project were published in Roscoe Braham's master's thesis, although the study is most often referenced from a now-classic volume published in 1949 by Byers and Braham. The most important finding came from radar. On the basis of radar measurements, they showed that most thunderstorms have well-defined regions of precipitation that are tens of miles in length and width, and these undergo well-defined life cycles. Byers and Braham used the biological term *cell* to describe these. They divided each cell's life cycle into three stages: *cumulus, mature,* and *dissipating* (Figs. 1.8, 1.9, 1.10, 1.11).

(a)

(b)

1.9 *Example of part of the life history of a thunderstorm over the Oklahoma Panhandle near Guymon on August 22, 1988, around 7:45 P.M. (a) the early mature stage; (b) when the anvil has become wider; and (c) the later mature stage. Under conditions of low vertical wind shear or when looking along the direction of the anvil-level wind, the anvil appears symmetrical.*

(c)

(a)

1.10 *Example of the life history of a thunderstorm on August 28, 1971, south of Miami, Florida: (a) the cumulus stage; (b) the mature stage, with the top of the large cloud tower hitting the tropopause; and (c) with the anvil well formed.*

(b)

(c)

1.11   *The dissipating stage of a thunderstorm on August 28, 1973, near Naples, Florida.*

During the first stage, cumulus clouds grow into towering mountains. The clouds can build as high as the tropopause, the transitional boundary between the troposphere and the stratosphere, 35,000–50,000 feet above the earth's surface.

Upward motions, or *updrafts*, on the order of 5 to 20 mph are common. Aircraft that clip towers of cumulus clouds will, at the least, experience nasty bumps as they hit the updraft. At worst, the aircraft may become dangerously out of control.

At high altitudes, where it is very cold, ice crystals form. When water vapor turns to liquid water or ice and cloud droplets freeze inside the clouds, latent heat energy is released, further enhancing the buoyancy of the growing clouds. But at the tropopause, above which the air is very stable, the buoyancy is quickly reduced to zero and the updraft dies.

The tropopause acts like a ceiling to the growing clouds. If there is no vertical wind shear, the top of the cloud, which is composed mainly of ice crystals, spreads out like a pancake to form what is called an *anvil* (Fig. 1.12). If there is vertical wind shear, the anvil may spread out in the *downshear* direction (Fig. 1.13), taking on a true anvil shape. Anvils, which cut down on surface heating during the day, may spread a hundred miles or

1.12 *A symmetrical-looking thunderstorm anvil viewed from the air over eastern Kansas near sunset on June 18, 1979.*

1.13 *Thunderstorm anvils spreading out in the downshear direction: (above) the world's narrowest anvil, Coral Gables, Florida, Sept. 1, 1972; (opposite page) anvil is seen streaming overhead, August 18, 1972, in Coral Gables.*

1.14 *A penetrating (or overshooting) top of a thunderstorm poking up above the tropopause on May 1, 1980, to the southeast of Norman, Oklahoma, at about 7:30 P.M.*

more downstream, heralding the approach of a storm. The thickness of an anvil and the sharpness in the appearance of its edge indicate to some extent the intensity of the updraft that collapsed into the anvil.

Some updrafts are so strong that they have enough kinetic energy to punch through the tropopause before they collapse. They form *a penetrating* or *overshooting top* (Fig. 1.14). When a penetrating top loses its bubbly shape and falls back down to the anvil level as a fibrous-looking cloud, it is sometimes referred to as *splashing cirrus* (Fig. 1.15).

Mamma (Figs. 1.16 and 1.17), or pouchlike extrusions at the base of the anvil, are often seen. Mamma appear to be smooth; when viewed up close, they may seem translucent (Fig. 1.18). Depending on the viewing angle, they may appear uniformly gray, deep blue, bright orange, or red. They are sometimes observed in bands that may be caused by upward and downward oscillations of air at the stable tropopause. At one time folklore had it that mamma turn into tornadoes, but this is unlikely since, among other reasons, mamma are usually at least fifteen thousand feet above the ground, and many tornadoes start at ground level.

Ice crystals beneath anvils may fall out and *sublimate,* or turn directly into water vapor. In this process heat is extracted from the air and it cools, perhaps enough to create negative, or downward buoyancy.

Eventually the cloud droplets in the updraft of the cloud's tower may grow into raindrops, which, like temperature, affect the cloud's density. The

1.15 An airborne view of pink-tinged "splashing" (or "jumping") cirrus at the top of a thunderstorm on June 8, 1990, over southwest Kansas about 9 P.M.

molecular weight of water vapor is lighter than that of the other gaseous constituents of the atmosphere, so water vapor slightly lowers the density of the air from what it would be if the air were completely dry. Liquid water, especially raindrops, has a much greater effect, in the opposite sense, because the density of water is a thousand times the density of air. Raindrops are heavy and detract significantly from buoyancy; they can become so heavy that the buoyancy disappears and rain begins to fall. If the vertical wind shear is weak, much of the rain falls out into the updraft. At this stage the thunderstorm is mature, and the cloud is a *cumulonimbus*—a cumulus cloud with rain.

The cumulonimbus, also known as a *cb*, needn't produce thunder and lightning, although sometimes it does. With its nearly vertical sides and

1.16 *Mamma underneath the anvil of a severe thunderstorm at sunset over southwestern Oklahoma and the Texas Panhandle on May 4, 1989, viewed to the west.*

1.17 *Mamma underneath the anvil of a thunderstorm, viewed to the northwest, in central Kansas on June 14, 1970.*

1.18   *A close-up view on May 26, 1991, at 7:36 P.M., from a NOAA aircraft, of mamma underneath the anvil of a complex of thunderstorms in northwestern Oklahoma, one of which had produced a tornado earlier. Note that the mamma appear to be translucent.*

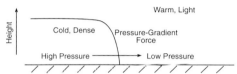

1.19   *How a density current works. The hydrostatic pressure on the cold side of the gust front (gust front boundary is marked by solid line) is higher than it is on the warm side. There is therefore a net pressure-gradient force that acts from the cold side to the warm side at the leading edge of the gust front.*

smooth anvil tops, cumulonimbus clouds often resemble magnificent works of art sculpted by the wind.

The falling rain may mix with unsaturated air and partially evaporate. As some of the rain, usually the smaller drops, evaporates, the air cools and loses buoyancy. The result is a *downdraft* driven by the cooling and the weight of the rain. When the downdraft hits the ground, it spreads out like a pancake, just as the updraft does when it hits the tropopause. The leading edge of the cool downdraft, marked by a shift in wind and an increase in gustiness, is called a *gust front*.

Behind the gust front the rain-laden air is cooler and heavier than the air in front of it. It moves like a *density current*, a term fluid dynamicists use to describe a current of air or water driven by a density difference (Fig. 1.19). Water flowing over a dam moves as a density current, and so perhaps does the Blob in the classic 1950s horror movie of the same name.

If the vertical shear in the storm is so weak that the gust front expands away from the cloud in all directions, blocking the upward flow of warm, moist air, the updraft dies out, and all that is left is an anvil with falling rain. This is the dissipating stage of a thunderstorm.

The entire life cycle of a Byers-Braham thunderstorm cell lasts about thirty to fifty minutes. But since ice crystals in the anvil sublimate relatively slowly, the anvil can persist long after the cell has dissipated, and the gust front can still kick up dust and sand. Anvil debris are sometimes referred to by weather observers as *orphan anvil*. Sandstorms created by the strong winds just behind the leading edge of a gust front are *haboobs* (Fig. 1.20).

In a downdraft, the straight-line winds (as opposed to wind spiraling in a vortex) may be intense enough to cause damage. Intense downdrafts

1.20 *A haboob near Winslow, Arizona, on August 3, 1978. Compare the shape of the leading wedge of airborne dust to the cold-air boundary depicted in Fig. 1.19.*

about five miles across are called *downbursts*, and small-scale downbursts are *microbursts*. Ted Fujita was a pioneer in the study of microbursts and their effect on aircraft that fly through them (Fig. 1.21). An aircraft that encounters a microburst while landing meets a strong headwind that gives the plane aerodynamic lift. The pilot must counter this increased lift by dropping the wing flaps enough to reduce lift. But at this point, pilot and plane are already in the middle of the microburst, beyond which the head-wind quickly turns into a tailwind, and the lift on the aircraft is suddenly reduced. The pilot must then extract all the power possible from the engines to avoid being rudely slammed into the ground.

Much of what we know today about microbursts, which are classified as either wet or dry (Fig. 1.22), came from field experiments conducted in the 1980s in Alabama and Colorado. Dry microbursts (Fig. 1.23) occur in an environment characteristic of the high plains of the United States—precip-itation falls from a relatively high cloud base into very dry air, where it rapidly evaporates, cooling the air enough so that it becomes even more

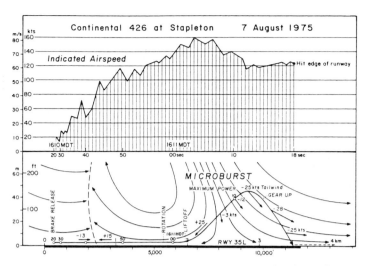

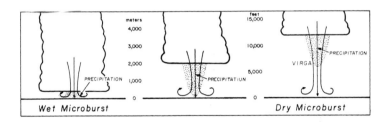

1.21 *The effect of a microburst on an aircraft. Flight path and indicated airspeed of Continental Airlines Flight 426 at Stapleton Airport, Denver, Colorado, on August 7, 1975. (From Fujita 1985.)*

1.22 *Schematic views of wet and dry microbursts. (From Fujita 1985.)*

1.23 *A dry microburst near Syracuse, Kansas, as viewed to the southwest from a NOAA aircraft on May 31, 1994, at 6:13 P.M. Note the precipitation falling at the far left, the gust probe on the aircraft seen at the lower left, and the half ring of blowing dust on the ground at the lower center of the photograph.*

1.24 *A wet microburst over Oklahoma City, Oklahoma, on July 26, 1978. This microburst produced very strong surface winds.*

negatively buoyant. Wet microbursts (Fig. 1.24) appear when dry air mixes with clouds bearing precipitation so that evaporative cooling induces negative buoyancy.

Downdrafts reaching the ground are almost always cold, but on rare occasions a negatively buoyant downdraft may encounter a layer of relatively cool air in which the negative buoyancy disappears. But the downdraft may be strong enough to make it to the ground before losing momentum. The result is that the air is compressed and warmed considerably by the time it reaches the ground. Such a phenomenon is called a *heatburst* (Fig. 1.25). Heatbursts near dissipating storms in the high plains have been known to increase the air temperature at night to as high as 100°F.

Thunderstorm cells begin when dry thermals, buoyant parcels of unsaturated air, reach the condensation level and remain buoyant. In the absence of dry thermals, the cells can be triggered if unsaturated air is lifted to its condensation level through various means. For example, air can be lifted up over a gust front until the condensation level is reached and clouds form. Such a cloud, in the absence of buoyancy, has a smooth, shelflike appearance (Fig. 1.26) and is called a *shelf cloud* or *arcus,* which means "bow" or "arch" in Latin. Shelf clouds are not to be confused with *roll clouds,* which

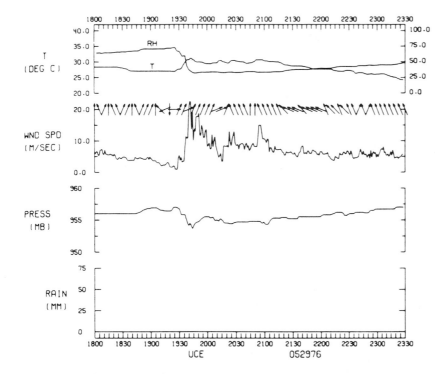

1.25 *Meteorological measurements in a heatburst on May 29, 1976, at Union City, Oklahoma. The heatburst occurred just after 1930 (CST). Note how the temperature and wind speed rose, while the pressure and relative humidity fell. The sinking air was drier; since it was also warmer and less dense than the surrounding air, the hydrostatic pressure decreased. (From Johnson 1983; courtesy of the American Meteorological Society. Copyright 1983.)*

1.26 *A shelf cloud over Norman, Oklahoma, on May 27, 1977, around 7 A.M., viewed from the east. The shelf cloud marked the leading edge of a mesocale convective system racing eastward across Oklahoma. Winds from the outflow blew out the window screens in the author's residence. The shelf cloud is produced as stable air is lifted over the low-level cold dome of air.*

1.27 *A roll cloud over Norman, Oklahoma, on June 21, 1978, at about 8:30 P.M., viewed from the west, looking to the northeast.*

are found behind a gust front (Fig. 1.27). Underneath the shelf cloud but ahead of the precipitation region, the scalloped cloud base overhead arches majestically (Fig. 1.28); this region has been called the whale's mouth, in honor of a memorable scene from the movie *Pinocchio* in which the textured inside of a whale's mouth is graphically seen.

If the condensation level is reached and the air becomes buoyant, a new cell may be initiated—which in turn triggers *secondary* cells along the edge of its gust front. The complex of cells produced by such a process is called a *multicell storm* (Figs. 1.29, 1.30) and is responsible for the relative longevity of a thunderstorm complex.

When gust fronts from a number of cells merge to produce one broad pool of cold air, the edge of the boundary of evaporatively cooled air is called an *outflow boundary*.

Based on the results of the Thunderstorm Project, research flights were undertaken around severe thunderstorms in the central United States in 1956 in the Tornado Research Airplane Project, or TRAP. Pilots who flew into squall lines in this project became known as the Rough Riders. In 1961, when the project had become the National Severe Storm Project (NSSP), investigators found that winds tended to flow around storms at many levels,

1.28 *The "whale's mouth" as seen behind the gust front of a thunderstorm on June 7, 1993, at 5:11 P.M., north of Howard, Kansas.*

as if the storms themselves were solid obstacles—which of course they aren't. In 1964 the Weather Radar Laboratory in Norman, Oklahoma, merged with the NSSP to become the National Severe Storms Laboratory.

Meteorologists refer to similar groups of thunderstorm cells that are organized on horizontal scales of fifty to a hundred miles or more as *mesoscale convective systems,* or *MCSs.* They tend to be nocturnal, having evolved from more isolated cells or lines of cells late in the afternoon or early in the evening, and they are responsible for much of the rain during the growing season in the midwestern United States. MCSs last for many hours, sometimes continuing through the night until dawn.

As thunderstorm cells evolve into MCSs, many become squall lines. Often a line structure develops with time, even if the squall line begins as an isolated cell or group of cells. The conglomeration of evaporatively cooled air masses from each cell eventually produces a large area of cold air at the surface, the leading edge of which forms a nearly straight boundary.

Often the most intense precipitation is along the leading edge of the squall line (the leading convective line), and there is a slackening of rainfall behind it in the transition zone, followed by a broad area of relatively uniform, heavy precipitation to the rear (Fig. 1.31). The latter is referred to as the trailing stratiform-precipitation area. The model of squall line MCSs depicted in Fig. 1.31 is valid for at least several hours after the first cells of line segments appear.

Squall lines with trailing stratiform precipitation areas sometimes evolve into bow-shaped structures (Fig. 1.32), dubbed "bow echoes" by Ted Fujita. Morris Weisman at the National Center for Atmospheric Research

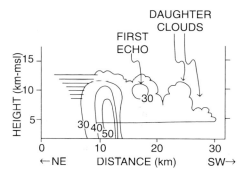

1.29 *An example of the behavior of a multicell storm. Vertical cross section from the northeast to the southwest of a typical eastward-moving multicell storm showing new cloud growth adjacent to older clouds. Height shown in kilometers above mean sea level (msl). New cells form to the southwest, develop radar-detectable precipitation aloft, and then a deep radar echo; the figure is in a sense a look at the different stages in the life of a thunderstorm cell. Cloud (scalloped area); radar reflectivity in dBZ (solid lines). (dBZ is an abbreviation for "decibels with respect to the logarithm of Z," where Z is a measure of the intensity of the precipitation; it is sufficient to know only that the dBZ scale is such that 20 dBZ represents light rain, while 50 dBZ represents heavy rain or even small hail.) (After Dennis et al. 1970; originally from Browning 1977. Adapted from Chisholm and Renick 1972 and Browning 1977. Courtesy of the American Meteorological Society. Copyright 1970 and 1977.)*

1.30 *Photograph of a multicell storm, near Denver, Colorado, on July 13, 1996, during the afternoon. New cloud growth is occurring to the south (right). Compare with the idealized depiction in Fig. 1.29. At the lower right a sloping (downward, from left to right) gust front boundary, which is moving to the right, is visible.*

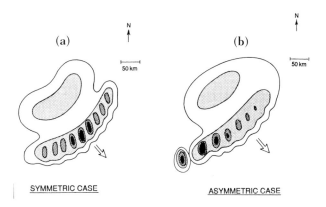

1.31 *Idealized radar depictions of a squall line having a leading convective line and trailing stratiform-precipitation area. Large vector indicates direction of motion of the squall-line system. Levels of shading denote increasing radar reflectivity, with the most intense values corresponding to convective cell cores. Horizontal scale and north arrow are as indicated: (a) "symmetric" type of squall-line system; and (b) "asymmetric" type of squall-line system. (From Houze et al. 1990; courtesy of the American Meteorological Society. Copyright 1990.)*

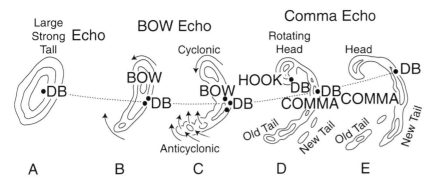

1.32 *The radar depiction of the formation of a bow echo as a function of times A–E (intervals between panels vary between 40 and 100 min). The locations of strong downbursts are labeled DB. "Bookend" vortices are indicated in C. (Adapted from Weisman 1993; after Fujita [1981]. Courtesy of the American Meteorological Society. Copyright 1993.)*

(NCAR) in Boulder, Colorado, used numerical simulations to explain their structure. Bow echoes are sometimes associated with damaging winds, tornadoes, and counterrotating vortices called *bookend vortices,* at the ends of the leading line. MCSs that last for long periods of time and produce damaging winds at least once every three hours are called *derechos*—from the Spanish for "straight ahead."

Vortices may also develop in the stratiform-precipitation region of squall lines. Ned Johnston from the University of Wisconsin at Madison first noted cyclonic swirls in the cloud tops associated with the remains of MCSs in satellite photographs. These vortices have nothing to do with tornadoes or other severe weather. But if they persist, they may be associated with a new round of thunderstorms the following afternoon or evening. Some theorize that over the tropical oceans, the vortices in the stratiform-precipitation area of MCSs can develop into hurricanes. (Hurricanes do not develop from MCS vortices over land because hurricanes require a warm ocean below to form.)

All the thunderstorms that Byers and Braham studied were ordinary garden-variety storms. The vertical shear of the wind was never too great, and they were brief and caused no damage. What about long-lived, damaging thunderstorms produced in an environment of great vertical wind shear?

Major advances in our understanding of severe thunderstorms came from British meteorologists Keith Browning and Frank Ludlam, and from American meteorologist Ralph Donaldson, who analyzed the radar data for two thunderstorms—one that spawned hail, the other, a tornado. The hailstorm occurred on July 9, 1959, in Wokingham, England (hardly a hotbed of severe thunderstorms), and the tornadic storm took place on May 4, 1961, near Geary, Oklahoma, about an hour's drive west of Oklahoma City.

Browning and his colleagues determined that the three-dimensional structure of the radar echo of these storms had a relatively long-lived col-

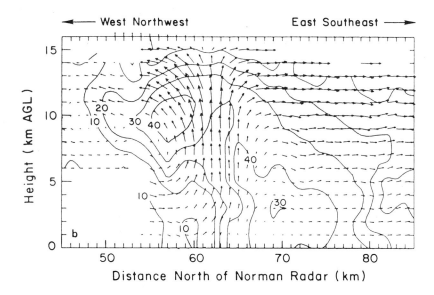

1.33 *A radar vault in a supercell in Central Oklahoma on April 26, 1984. Vertical cross section at 2345 UTC of radar reflectivity (solid contours in dBZ) and wind field (vectors). Vault is located about 63 km north (and 13 km west) of the Norman radar, 8–11 km above the ground. Wind Scale: Distance between tick marks represents 30 m/sec. (From Bluestein and Woodall, 1990; courtesy of the American Meteorological Society. Copyright 1990.)*

umn of relatively weak radar echoes, which they termed the *vault* (Fig. 1.33). They called a horizontal cross-section through a vault a *weak-echo region*, or *WER*. Aloft, the weak-echo region, which is often enclosed, is called a *bounded weak-echo region*, or *BWER*. An *echo-free hole* was also noted. They hypothesized that the vault owed its existence to a *persistent updraft*, that is, air was moving upward so quickly that by the time raindrops formed, the rain was near the top of the storm.

They also emphasized that the storms seemed to be steady, remaining unchanged for about an hour or more. In contrast, ordinary thunderstorm cells dump their rain quickly and are over in twenty to thirty minutes. Furthermore, while most thunderstorms seem to be steered along by the mean wind (the wind vector averaged vertically with respect to the mass of the air) in the troposphere, the Oklahoma and England storms moved to the right of the mean wind. If the winds were carrying rain, the researchers deduced, then the movement to the right must be because new cell growth was continuous adjacent and to the right of the older cell, a process they termed *continuous propagation*. Propagation is what is responsible for wave motion at the surface of a body of water; the crest of a wave hitting the shore appears to move along, but in reality the water surface is actually only bobbing up and down in place.

In 1964 Browning coined the term *supercell* for a large, persistent, right-moving, damaging thunderstorm that endures for an hour or more. Most supercell thunderstorms produce large hail. And supercells are prolific breeders of tornadoes. Why do supercells behave differently from the

Byers-Braham cells of ordinary thunderstorms? A key factor appears to be the amount of vertical wind shear.

Supercells are the most prolific producers of large hail (three quarters of an inch in diameter or larger), but large hail can also be produced in ordinary cells. Strong updrafts are necessary to keep heavy chunks of ice suspended long enough for them to accrete cloud water and rain, which then freeze. When hailstones are cut open, they often look like tree rings (Fig. 1.34). Concentric shells of different shades and textures of white suggest that the stones may have been bounced up and down in updrafts several times before finally falling to the ground. The trajectory a hailstone takes through a storm determines how large it will get and where it will fall out. An analysis of size is noteworthy. Storm spotters rarely report hail size according to measurement. Most often they compare it to common objects: grapefruit, softballs, baseballs, golfballs, Ping Pong balls, hen's eggs, quarters, nickels, dimes, or peas.

A phenomenon anecdotally related to hail is the *green thunderstorm*. Regions of precipitation behind gust fronts sometimes appear to be brilliant green. Atmospheric physicists Craig Bohren and Alistair Fraser have proposed two hypotheses to explain these storms, neither of which requires hail. One involves a dark backdrop against which red light from the setting sun is cast; the other involves absorption of red light through water or ice. Atmospheric physicist Frank Gallagher has also been measuring the light spectrum in green thunderstorms from storm-intercept vehicles and from aircraft, and has verified their greenness quantitatively. I have observed many green thunderstorms and have found them not necessarily associated with hail. In fact, in my experience, the greenish light is associated with heavy rain or heavy rain mixed with small (pea-size) hail, or just a lot of small hail. Greenish colors are also seen adjacent to wall clouds and as linear features above shelf clouds for reasons not yet explained completely, though it may be related to the thickness of the cloud.

The optical effects associated with hail can be intriguing, but hailfall can also create interesting patterns on the ground. For example, narrow swaths of hail reveal the limited extent of hailfall regions within thunderstorms. Sometimes a foot or more of small hail can accumulate or be swept into piles, forcing snowplows out of storage to clear the roads. Hail can transform an idyllic spring day into a winter wonderland in a matter of minutes. An inch or more of hail will easily cool the surrounding air, causing *hail fog* to form. Though beautiful to look at, hail fog can be very hazardous, especially to drivers.

In the 1950s and 1960s severe thunderstorms were termed *severe local storms* to distinguish them from the larger-scale, nonlocal storms, such as *extratropical cyclones* and *hurricanes*, which span distances of hundreds and thousands of miles. A severe local storm was considered to cover an area only ten miles by ten miles or so

To a forecaster today at the Storm Prediction Center in Norman, Oklahoma, "severe" means a tornado or a storm with hail in excess of three fourths of an inch in diameter or wind gusts greater than 55 mph (around gale force or greater).

1.34 *Large hailstones cut open, revealing concentric rings, on May 16, 1991, near El Dorado, Kansas.*

As defined by the U.S. National Weather Service, the area of a *severe-thunderstorm watch* is usually a rectangular- or parallelogram-shaped region (sometimes referred to as a box) about two hundred or three hundred miles by one hundred miles. A watch means only that a severe storm is anticipated. If one does show up on radar, a warning is issued for a much smaller area, usually on the scale of a county or two. Warnings usually last for less than an hour, while watches may continue for more than three hours.

I like to think that "severe" means capable of inflicting damage. A sturdy structure can easily survive wind gusts of 70 mph, but a trailer can be badly damaged by lower wind speeds. The National Weather Service does not consider a thunderstorm severe if it produces weak winds and no hail. Yet the lightning and flash flooding associated with a so-called weak thunderstorm can kill people. Knee-deep drifts of pea-sized, "nonsevere" hailstones seriously impair driving. And some tornadoes are spawned by ordinary thunderstorms.

The area of a tornado watch is similar to that of a severe-thunderstorm watch. Tornadoes are expected to develop within a tornado-watch area but are not anticipated in a severe-thunderstorm-watch area. Yet it is not unheard of for tornadoes to touch down within a severe-thunderstorm-watch area—and sometimes no tornadoes show up within a tornado-watch area. Forecasters trying to determine whether to issue a tornado watch or a severe-thunderstorm watch make their decisions subjectively.

Severe-weather forecasts were first issued formally in 1952 by the Severe Local Storms Forecasting Unit (SELS) of the U.S. Weather Bureau in Washington, D.C. In 1954 SELS moved to Kansas City. In 1966 it was renamed the National Severe Storms Forecast Center (NSSFC). In 1995 NSSFC was renamed the Storm Prediction Center (SPC). As of this writing, the SPC is in Norman, Oklahoma.

Light and Color

When I left graduate school, my notion of a severe thunderstorm was that of most others—a menacing, dark sky and unusual colors. Former NSSL director Ed Kessler has remarked that a dark cloud base may indicate a strong updraft and implies that sunlight is significantly attenuated by cloud and rain droplets suspended aloft. Before precipitation starts, the cloud tends to darken. When rain does fall, it scavenges the cloud droplets and smaller drops in its descent, and more sunlight shines through the clouds. If precipitation is very heavy, the cloud tends to be relatively bright. Severity, then, is not necessarily related to darkness. Some storms produce tornadoes and large hail in the absence of a dark sky. Unusual colors are sometimes seen in severe storms. Dust blowing to the west may tint the sky reddish or orange; bright green clouds could indicate an updraft so strong that not much cloud material is present and quite a bit of sunlight is filtering through.

2

Catching
Real Storms

I wield the flail of the lashing hail,
And whiten the green plains under,
And then again I dissolve it in rain
And laugh as I pass in thunder.
                    —*Shelley*

THE SKY IS ONE OF nature's art
museums. Its exhibits are constantly changing—and sometimes they can
literally blow the viewer away.

Consider the tornado or the thunderstorm. The birth of these natural
objets d'art begins with the production of buoyant blobs of air caused by a
large temperature difference between a layer of air close to the ground and
the air above it. This results in the formation of convective clouds.

In small-scale atmospheric convection—from the Latin *converhere*, "to
bring together"—the atmosphere moves in response to density differences
brought about by differential heating. The result of atmospheric convection is a redistribution of heat, and its manifestation is a continuous source
of buoyant air in the form of a plume (Fig. 2.1) or isolated bubbles.

Buoyancy, the same physical process responsible for making ice float in
a glass of water, causes the bubbles or plume of air to rise. There are several forces acting on a buoyant parcel of air. Gravity tries to push it downward; the pressure-gradient force, which acts upward from relatively high
pressure (with respect to the pressure away from the parcel) to relatively
low pressure, tries to push it upward (Fig. 2.2). When the force of gravity is

35

(a)

2.1 *The explosive nature of convection: (a) a convective plume induced by an industrial heat source in Hayden, Colorado, on January 12, 1995; it is not uncommon for industrial heat sources such as power plants to trigger, during the winter, a convective cloud that locally produces snowfall; (b) an airborne view of a convective plume induced by an erupting volcano west of Mihara Yama, Japan, at 3:59 P.M. on July 14, 1987; and (c) a "pyrocumulus" induced by a forest fire in Boulder, Colorado, on July 9, 1989, at about 3:15 P.M.*

(b)

(c)

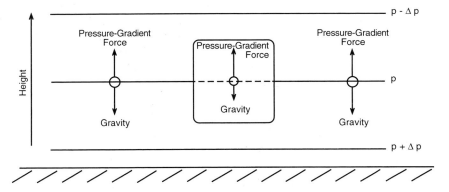

2.2 *An idealized depiction of the nature of buoyancy. Solid lines represent isobars denoted by p, where p+Δp is greater than p, which in turn is greater than p-Δp; the dashed line represents an isobar inside the buoyant blob of air (whose density is less than that in the air surrounding the blob). The enclosed thick solid line represents the vertical cross-section of the buoyant blob (in this case a cube) of air. Outside the buoyant blob the atmosphere is hydrostatic. The pressure field inside the buoyant blob is the same as it is outside the buoyant blob. There is a net upward force (buoyancy) in the blob of less dense air because the upward-directed pressure gradient force is greater than the downward-directed force of gravity.*

exactly counterbalanced by the pressure gradient force, the parcel of air will remain suspended, and the atmosphere is said to be *hydrostatic.*

The pressure at any level in a hydrostatic atmosphere is equal to the weight of the air above it. In other words, pressure decreases with height, because the higher you go, the less air there is above you. Furthermore, air, unlike water, is compressible. Thus the air near the ground, with more weight above it, is squashed or compressed relative to the air aloft. Density therefore decreases with height. Although the acceleration of gravity within the earth's atmosphere is essentially constant, the force of gravity decreases with height.

If a bubble of air is less dense than the air around it, the force of gravity inside the bubble is less than it is outside. If the pressure is the same both inside and outside the air bubble, then the upward-directed pressure gradient force is greater than the force of gravity and there is an imbalance. Thus, the air parcel would have a net upward-directed force, the buoyancy force.

A bubble of air lighter than its surroundings is buoyant and pushes upward. If it is heavier than its surroundings, it is negatively buoyant and accelerates downward. Birds and glider pilots often make use of thermals—buoyant pockets of unsaturated air—to soar.

As a buoyant air bubble rises as an updraft, it moves into an environment of lower pressure, where it expands and loses energy in the form of heat. Conversely, a negatively buoyant pocket of air falls as a downdraft into a region of higher pressure and is compressed, gaining energy in the form of heat.

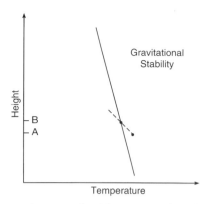

2.3 *An example of (gravitational) stability. The solid line represents the vertical variation of temperature in the atmosphere; note that in this case the temperature decreases with height. The dashed line represents the temperature a buoyant air bubble at height A would have if it were allowed to accelerate upward and if it did not mix with its surroundings. After it reaches B, it becomes colder than its surroundings (i.e., negatively buoyant). It decelerates and reverses direction, falling back to B. If it overshoots B, it becomes buoyant and bounces back up again, and so on.*

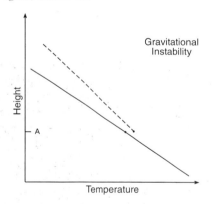

2.4 *An example of (gravitational) instability. The solid line represents the vertical variation of temperature in the atmosphere; note that in this case the temperature also decreases with height, as it does in Fig. 2.3. However, here the temperature decreases more rapidly than it does in Fig. 2.3. The dashed line represents the temperature an air bubble at height A would have if it were allowed to accelerate upward and if it did not mix with its surroundings. Above A it would remain warmer than its surroundings, buoyant, and continue to accelerate upward.*

The atmosphere obeys the ideal gas law, which states that pressure is proportional to both density and temperature. At a given pressure, density goes down as the temperature goes up, and density increases as the temperature decreases.

At a fixed pressure, warm air is lighter than cold air. A rising bubble of dry air cools at the rate of about 18°F with every 3,280 feet of height. The rate of temperature decrease with height is called the *lapse rate*. If a rising buoyant air parcel cools at a rate greater than the lapse rate (Fig. 2.3), the air will eventually cool so much that it will become denser than its surroundings and be pushed downward. If there were no friction, the air would warm at a rate greater than that of the surrounding environment, become buoyant again, and bounce up and down like a wave. Such a scenario is one of gravitational stability; the wave motions are called *buoyancy oscillations*. The typical period of oscillation is about ten minutes. In a *stable* atmosphere, the lapse rate is *less* than 18°F per 3,280 feet.

If a rising air parcel cools at a rate less than the lapse rate (Fig. 2.4), it will remain buoyant and continue to be pushed upward. Such a case is one of gravitational instability. The atmosphere is said to be *absolutely unstable* when the lapse rate *exceeds* 18°F per 3,280 feet. Such a state is typical near the ground on a hot, sunny day when the ground temperature is much warmer than the air just above it, or during the winter when very cold air flows over relatively warm water, such as frigid Arctic air flowing out over the Gulf Stream.

Now let us add water vapor to the rising buoyant air bubble. As it ascends, the air may cool so much that it becomes saturated. When the water vapor condenses, latent heat is released and absorbed by the bubble. A saturated rising bubble of air cools at a rate of approximately 12°F for every 3,280 feet of height (rather than 18°F for every 3,280 feet if it were unsaturated), owing to the addition of latent heat. At higher altitudes, where pressures are lower, the bubble cools more rapidly, eventually cooling at the same rate as dry air, because the amount of water vapor that can be suspended in the air decreases with height, owing to the decrease in temperature and pressure. If the atmosphere's lapse rate is less than 18°F per 3,280 feet, but greater than about 12°F per 3,280 feet, then the atmosphere is said to be *conditionally unstable*, that is, stable if unsaturated and unstable if saturated. It follows that an environment that can support convective clouds is relatively warm and moist at low levels, and cold at upper levels.

In a convective cloud, water vapor condenses on the surface of hygroscopic nuclei such as salt or dust particles. Clouds are sculpted by the air flowing through them, often resembling cauliflower (Fig. 2.5). The English pharmacist and amateur scientist Luke Howard was responsible for naming, in 1802, small convective clouds *cumulus*—Latin for "heap."

The saturated bubble of air is now visible as a cumulus cloud; the air outside it is unsaturated and cooler. Air at the edges of the cloud mixes with some of the cloud droplets and some of the air from within the cloud; from above, some of the ambient air is drawn down into it. The net result is a less buoyant cloud. Then, as cloud droplets evaporate into the surrounding unsaturated air and the cooler ambient air is drawn into the

2.5 *A comparison of (a) a saturated, buoyant, convective-cloud top with (b) a head of cauliflower. The photograph in (a) was taken from an aircraft before landing to the north at Denver during the midafternoon of June 6, 1990, just hours before a powerful tornado roared through Limon, to the east. Stare at (or view time-lapse movies or videos of) the top edge of a vigorously growing cumulus cloud or towering cumulus to visualize the turbulent motions at the edge of the bubble. Any resemblance between the buoyant bubble and a head of cauliflower or a human brain (not shown) is purely coincidental.*

(a)

(b)

cloud, the cloud cools. In fact, it may cool so much that it loses its buoy-ancy and dissipates.

The rate at which cooler, unsaturated air is drawn into (or *entrained*) by a cloud depends on the width of the cloud. In wide clouds, the core of the buoyant updraft is protected from the cooler air outside the cloud; in narrow clouds, the core is more vulnerable. The wider the cloud, then, the more likely it is to rise.

Cloud Display

Shallow or small cumulus clouds come in a number of varieties. The most feeble cumulus form when rising buoyant air bubbles hit a stable layer, a layer marked by an increase in temperature with height or only a slight decrease in temperature with height. These clouds look as if their tops are flat (Fig. 2.6). They are called *stratocumulus—strato-* from the Latin *sternere*, "to spread."

Stratocumulus are often found in low-level air masses that feed severe thunderstorms. For example, over the Gulf of Mexico, evaporation from warm surface waters induces the formation of many saturated convective bubbles. If there is a layer of warm air above, the bubbles cannot rise any farther and so cannot mix with the drier air above; as a result, a warm moist air mass is produced near the surface.

If this moist Gulf air were to be transported (*advected*, in meteorologi-cal parlance) northward, and warmer dry air from higher terrain off to the west in New Mexico and to the southwest in Mexico flowed over the moist

2.6   (*left*) *Stratocumulus clouds viewed from high above over the Pacific Ocean on November 5, 1995, somewhere between Los Angeles and Hong Kong.*

2.7   (*right*) *A checkerboard of fair-weather cumulus clouds viewed from above near the Oklahoma–Texas border along the Red River, on June 24, 1984. Note how the clouds exhibit more vertical development than the stratocumulus shown in Fig. 2.6.*

2.8 *Towering cumulus clouds in the absence of strong vertical wind shear: (left) September 5, 1984, in Key Biscayne, Florida; (right) September 1, 1972, in Coral Gables, Florida.*

air, then the moisture would be constrained to remain near the ground because the air is so stable. Thus very muggy air can flow from the Gulf northward into the Great Plains of the United States while maintaining its moisture content. The stable layer of air above the moist air mass is called a *cap*, or lid, because it prevents low-level moisture from rising any farther. When the air above this cap is conditionally unstable and has a lot of potential buoyant energy, the atmosphere is said to have a loaded-gun structure: Deep convection is constrained initially, but when it begins, it does so explosively. The loaded-gun structure is also known as the Miller Type I sounding, after Captain Miller at Tinker Air Force Base near Oklahoma City, who in the early 1950s first identified it with severe weather.

A step up in vertical development from the stratocumulus are the *fair-weather cumulus*, or *cumulus humilis*—Latin for "humble" (Fig. 2.7). When cumulus clouds become very tall and narrow, they are called *towering cumulus*, or *congestus*, from the Latin *congerere*, "to pile up" (Fig. 2.8). Some like to call the narrowest cumulus congestus "turkey towers" because

2.9 *Towering cumulus clouds leaning over in response to strong vertical wind shear on April 27, 1983, viewed to the north in central Oklahoma, along a dryline. Cloud tops may break away to become "orphan anvils."*

they resemble the necks of turkeys. The bases of towering cumulus, also called TCU, are usually flat because the level at which condensation forms is relatively uniform. When towering cumulus grow in an environment in which the winds near the top of the cloud are blowing at a different speed or from a different direction than the winds below, that is, where vertical wind shear is significant, then the tower may start to lean, looking as if it might topple over (Fig. 2.9).

Shallow cumulus or towering cumulus are sometimes organized into lines, or "cloud streets" (Fig. 2.10), or each may be spread out like the squares on a checkerboard (see Fig. 2.7).

Cumulus convection can also be initiated high in the atmosphere if the air is lifted by means other than a thermal. For example, the air associated with large-scale disturbances high in the atmosphere may be ascending at a relatively slow rate (around a mile per day) over a broad area (around a thousand miles by a thousand miles). When a parcel of air in the troposphere reaches its condensation level, buoyancy is realized in turret-shaped clouds called *altocumulus castellanus*—castles in the sky, if you will. Waves of rising and falling motion in the atmosphere may trigger their formation. These clouds often appear in lines, perhaps because the pattern of rising air that forms them is linear. Their top edges usually are not so sharply defined as those of ground-based cumulus. When the castellanus lose their buoyancy, they tend to resemble tufts of wool (Fig. 2.11). Clouds displaying this feature are of the *floccus* variety—after the Latin for "flock of wool."

2.10 *Cumulus cloud streets as seen from high above over the Pacific Ocean on November 5, 1995, somewhere between Los Angeles and Hong Kong.*

2.11 *Altocumulus castellanus floccus over Norman, Oklahoma, on October 4, 1976.*

The appearance of altocumulus castellanus is visual evidence of conditionally unstable air in the troposphere. A thunderstorm may grow from altocumulus castellanus (also called ACCAS), but I've never known them to spawn a tornado.

The First Storm Chasers

Before the 1970s, most recorded observations of tornadoes were serendipitous. Useful and intriguing information had been learned about tornadoes and their parent storms, but it had come from chance encounters with tornadoes within the range of radar, surface instruments, and people. As the research meteorologist Louis J. Battan noted in the preface to his 1961 book on violent storms, the atmosphere itself is in large part to blame. Battan wrote:

> Regardless of the types of experiments a scientist may perform in his laboratory, regardless of the mathematical calculations he may make with his electronic computer, in the end he must go out and make measurements in the atmosphere. . . . At this stage of the investigation real difficulties frequently arise. One obvious problem is that of finding the type of weather disturbance he wants to study at the right time and in the right place. . . . How would you go about making measurements in a tornado knowing that one would probably form in the state of Kansas, but that it would be only 300 feet across and would last for two minutes? Then, bearing in mind that winds of several hundreds of miles per hour would be present in the tornado, what types of instruments would you use? . . . The atmosphere will not stand still and be measured.

Roger Jensen, who might be the oldest storm chaser alive, was born in 1933 and began chasing in 1953 in the Dakotas and Minnesota, but not for scientific purposes. The late Neil Ward, at the National Severe Storms Laboratory, went out on his own storm chases as early as 1961 and met with some success. Dave Hoadley, another early storm chaser, began as a teenager in Bismarck, North Dakota, during the summer of 1956. He chased twenty thousand miles before seeing his first tornado. With an education in the liberal arts and no formal meteorological training, Dave devised a method of weather data analysis using his own symbols for such features as cold fronts. His outstanding cloud photographs, which I first saw at a conference in Omaha in 1977, were an inspiration to me to study severe storms, as I am sure they have been to many others.

By the 1970s, however, it had become increasingly apparent to meteorologists that rather than wait passively for the storms that might strike nearby, it would be worthwhile to take a more offensive stance and head out into the field to meet the storms. The first systematic attempt to intercept severe thunderstorms and tornadoes for scientific purposes began in the early 1970s at NSSL. The Tornado Intercept Project, directed by meteorologists Joe Golden and Bruce Morgan, focused on central Oklahoma. Some students from the University of Oklahoma also went out after storms on their own, and if they weren't officially part of the project, they at least had the tacit support of Professor Yoshi Sasaki, a pioneer in the development of techniques for analyzing meteorological data. The first tornado was intercepted on April 30, 1972, in western Oklahoma. On numerous occasions, the storm chasers confirmed the cloud architecture found by Fujita more than a decade before in the Fargo storm.

From this project and from subsequent storm chasing, a visual model of a tornadic supercell was developed (Figs. 2.12, 2.13) and used by storm spotters to help warn the public of a tornado in the making. The model is not meant to represent any particular storm but rather the general features found in a class of storms. According to the model, tornadoes tend to be located under a wall cloud, a relatively precipitation-free (at least visually) cloud base that extends below the neighboring cloud base and is located under a rapidly growing convective cloud. The right-hand side, when viewed from the front with respect to its motion, sometimes resembles a vertical wall (Fig. 2.14). The practical importance of the wall cloud is its association with tornadoes and its nearness to a region of large hail (see Fig. 2.12). Sometimes seen stretching to the right of a wall cloud is a tail cloud that extends into an area of precipitation, from which cloud fragments move toward the wall cloud.

Some wall clouds that I have watched begin as *scud*. Scud clouds, which are relatively amorphous, form below the preexisting cloud base when air with above-average humidity rises. The scud then ascend and attach themselves to the cloud base, effectively lowering it. Such clouds

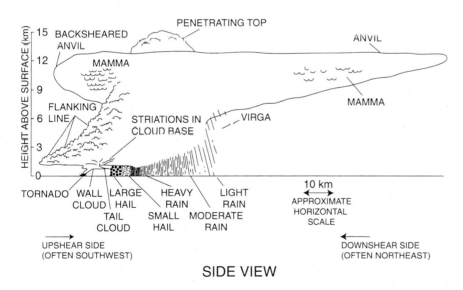

2.12 *The visual model of the mature phase of a tornadic supercell in the Northern Hemisphere as viewed from a vantage point approximately ahead and to the right of the storm's track. Prominent features include the wall cloud, tail cloud, tornado, flanking line, backsheared anvil, penetrating top, anvil, mamma, virga, and precipitation areas. (After a figure by A. Moller, adapted from Bluestein and Parks 1983; courtesy of the American Meteorological Society. Copyright 1983.)*

2.13 *Photograph of a tornadic supercell near Ulysses, Kansas, in the southwestern part of the state, on May 26, 1991, at approximately 6:27 P.M., viewed to the west from a NOAA aircraft that was racing toward the storm. Note the remarkable similarity to the idealized visual model of a tornadic supercell seen in Fig. 2.12.*

2.14 *Wall clouds (a, right) viewed from an NOAA aircraft, from the southeast, looking to the northwest on May 26, 1991, between 6:30 and 6:36 P.M., in southwest Kansas; note the flanking line on the lower left; (b, bottom right) viewed to the northwest, from a location north of Munday, Texas on May 27, 1985 at around 8 P.M. (c, opposite page) (3-part panel) viewed to the west, south of Canyon, Texas, at around 7:51–7:53 P.M. on May 26, 1978; the tail cloud was moving very rapidly from right to left (i.e., from the north to the south), and hence appears slightly blurred; the location of the author-photographer was, ironically, by a cemetery; in the middle portion of the triptych, the wall cloud is so low it is almost touching the ground; (d, following page) viewed from the west, looking to the east, near Burbank, Oklahoma, on May 11, 1978, at 6:23 P.M.; there is no tail cloud.*

(a)

(b)

(c)

(**d**)

are visual evidence of more humid air from an adjacent region of precipitation; the increase in humidity is caused by the evaporation of some of the precipitation. This was supported in the early 1980s using numerical simulations of supercells. Tail clouds and roll clouds (see Fig. 1.27) may be thought of as a line of scud.

Some of the air entering a wall cloud is also cooler than the ambient air as a result of evaporation. When lifted under the cloud, this cooler air is responsible for the smooth, laminar, and sometimes striated appearance of the sides of the wall cloud. When the cooler air is lifted, it is less likely to be buoyant near the cloud base; in other words, it remains stable for some distance above the cloud base as it is being sucked up. Eventually the cloud becomes buoyant. Clouds produced when air is forced upward in a stable environment have a smooth appearance; an example is the *altocumulus lenticularis*, or lens-shaped wave cloud. Wave clouds are often seen when stable air is forced up to its condensation level over mountains or to their lee.

Wall clouds in supercells usually appear to rotate very slowly in a cyclonic manner (counterclockwise in the Northern Hemisphere) and tend to persist for ten to twenty minutes or more. Lowered cloud bases can also be seen in multicell complexes when relatively humid air from an adjacent cell is taken into a new updraft of a growing cell. But the lowered cloud base is not likely to rotate—precipitation will fall from it, and the wall cloud soon disappears. While supercell wall clouds rotate and persist at least as long as the Byers-Braham cells, multicell wall clouds are short-lived and don't rotate.

Wall clouds sometimes deform into horseshoe-shaped structures, with a slot of clear air prominent in the middle (Fig. 2.15). On rare occasions, wall clouds have eyes like those of hurricanes (Fig. 2.16), and some have

2.15 *Wall cloud deformed into a horseshoe shape on June 20, 1979, at around 8 P.M. southwest of Norman, Oklahoma; viewed from the southeast, looking to the northwest.*

2.16 *Wall cloud with an eye near Sedan, Kansas, on June 19, 1981, at 7:48 P.M.*

2.17   *Wall clouds viewed from the east, looking to the west, over the Texas Panhandle on May 26, 1978, at 6:30 P.M., north of Nazareth. Tail clouds are seen feeding into each of the two wall clouds from the right; each wall cloud is rotating in a cyclonic fashion about its own axis and also about the other wall cloud.*

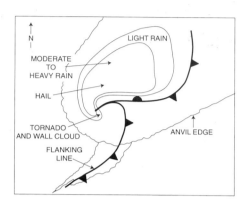

2.18   *Idealized model of the distribution of precipitation in a supercell in the Northern Hemisphere. (Compare with the visual features shown in Fig. 2.12; the viewer is located at the lower right edge of the figure to see the view shown in Fig. 2.12.) The wall cloud and tornado are located at the intersection of the leading edge of the rear-flank downdraft (cold-front symbol) and the edge of the forward-flank downdraft (warm-front symbol). In this example the storm is moving toward the northeast. When anticyclonic tornadoes do occur, they tend to be found along the flanking line gust front, several kilometers from the intersection of the rear-flank downdraft and the forward-flank downdraft. (Based on a figure from C. Doswell.)*

more than one vortex (Fig. 2.17). Tornadoes tend to occur on the southern edges of these clouds and rotate around them, ending up at the northwest tip of the horseshoe.

With respect to storm motion, the rear side of a supercell usually contains a flanking line of cumulus towers (Figs. 2.12, 2.18). Commonly stair-like in appearance (Fig. 2.19), this line may also be tilted. The projection of the line onto a horizontal plane sometimes arcs into a bow shape. It is often bright behind the flanking line. The wall cloud is usually found underneath the area where the flanking line intersects the tallest-growing tower in the storm.

In a supercell, the updrafts are often strong enough to produce penetrating tops (cloud material that pushes above the stable troposphere) and a backsheared anvil (Fig. 2.20). The latter feature indicates that some of the air being squirted in the main updraft at the tropopause level has enough kinetic energy to flow back against the upper-level winds, leaving some anvil material behind. Sometimes it is even strong enough to curl back upon itself. Both the backsheared anvil and the major portion of the anvil downwind of the storm often have mamma hanging from them. Sometimes the cloud base under a main updraft is flat, parallel to the ground, and very dark.

An observer in the path of a storm (or a stormchaser approaching from the downshear side) would first see the hard, crisped-edge anvil encroach overhead in a menacing manner, like the archetype cloaked villain. *Virga*, precipitation that falls out but evaporates completely before reaching the ground, is first seen under the anvil. On the horizon it is dark and light rain begins to fall, giving way to moderate and then heavy rain.

2.19 The "stairstep" appearance of the flanking line of a supercell, viewed to the east, on April 7, 1980, east of Eufaula, Oklahoma.

2.20 An anvil-cloud top that is curling back upon itself as it explodes upward near Archer City, Texas, to the northeast, on May 29, 1994, at 4:34 P.M.; viewed from the south, looking to the north, from an NOAA aircraft.

2.21 *Cloud-to-ground lightning just behind the leading edge of a thunderstorm in Norman, Oklahoma, on August 21, 1979.*

Cloud-to-ground lightning (Fig. 2.21) occasionally strikes beneath the storm. Lightning flashes have been observed everywhere within tornadic thunderstorms. Sometimes so many occur so quickly that at night the cloud appears continually illuminated against the dark sky. Highly branched lightning moves more slowly, spreading more than fifty miles along the base of the anvil (Fig. 2.22). This lightning, sometimes referred to as "anvil crawlers" or "spider lightning," can be spectacular in mesoscale convection systems, which have expansive anvils.

How storms become electrified enough to produce lightning is not completely understood. It appears that a mixture of ice crystals, super-cooled but not frozen water droplets, and strong convective updrafts are necessary. Although transfers of both negative and positive charges to the ground have been documented, most cloud-to-ground flashes transfer a negative charge. The location of lightning is determined from detection systems that make use of the electromagnetic noise radiated by the lightning—the noise you can hear as static on an AM radio.

Soon small hail may begin to mix in with the rain, perhaps reaching marble size. As the rain slacks off, the hail may grow to golfball or baseball size. At this stage, the hail is not a constant barrage, but a sporadic fall of large stones. Visibility increases rapidly as the density of the hail curtain decreases. Suddenly the wall cloud appears, and a tornado might be close by. Tim Marshall, a veteran storm chaser, refers to this potentially tornadic region under the wall cloud as the "bear's cage."

This visual model of the tornadic supercell storm has been used by storm chasers to position themselves safely in view of actual and potential tornadoes. The best place to be is a few miles ahead *and to the right* of the track of the wall cloud. In this location, a tornado will not usually threaten the observer, who will be able to see it clearly—and use a telephoto lens to get a close-up view. Of course, there are exceptions to the rule, especially if the structure of the storm deviates from this model or rapidly evolves into another type of storm.

The National Weather Service and even some television stations train amateur spotters to use the visual model for short-term forecasting of where within a storm a tornado might occur. But I must admit that I have never been able to look at a storm from a distance and identify it as one likely to produce a tornado in the next twenty minutes or so. Nor have I been able to look at a satellite image of storms and discriminate a tornadic storm from a nontornadic one. This really amounts to judging a book by its cover. Apparently what produces tornadoes may not be revealed at high levels, especially if the anvil blocks out the motion of the cloud features below or if the processes responsible for tornado production occur near the ground, below the clouds in the parent storm.

2.22 *Lightning branches in the anvil of a severe thunderstorm south of Norman, Oklahoma, on May 15, 1977.*

Chasing Waterspouts

In the late 1960s a parallel storm chase was underway in the Florida Keys. In this case, the prey was waterspouts, which are tornadoes over water (Figs 2.23, 2.24). They often occur under growing towering cumulus clouds that have not yet reached the cumulonimbus stage. Usually they are relatively weak (Fig. 2.23). But a few are strong tornadoes in severe thunderstorms that have moved or formed (Fig. 2.24) over water. And occasionally a waterspout makes landfall and inflicts damage.

In the Florida Keys waterspouts tend to form on days of "undisturbed weather," when large or severe thunderstorms are not expected. Many waterspouts are only about thirty feet across (see Fig. 2.23); most tornadoes are at least ten times wider. Boats have navigated through waterspouts and aircraft have skirted or clipped their condensation funnels without serious mishap. But waterspouts have been reported to have sucked fish and seaweed out of the water and rained them out over the land. Waterspouts have also been observed during volcanic eruptions on islands. Just south of Iceland, dozens of waterspouts formed downwind from the island volcano Surtsey during a major eruption in 1963. Many formed under the sloping plume (cloud) emanating from Surtsey.

Meteorologist Joe Golden, while still a graduate student, documented the life cycle of waterspouts. Golden's field study was apparently motivated

2.23   *A typical waterspout over the Florida Keys. This waterspout on August 24, 1993, at 12:01 P.M. is viewed from a NOAA helicopter during a project conducted jointly with the National Geographic Society. Ocean spray is visible above the sea surface, while a diffuse condensation funnel is visible aloft.*

2.24   *A large waterspout to the east of Key Biscayne, Florida, during the afternoon of May 28, 1975. The author took this photograph from his hotel-room balcony while attending his first professional conference, just before his presentation. The palm trees seen in the photograph are no longer there, thanks to Hurricane Andrew over fifteen years later.*

by a chance aerial encounter with a waterspout, and during the summer of 1969, he observed more than a hundred waterspouts in the Florida Keys. While I was a graduate student at MIT, I spent summers at the National Hurricane Research Laboratory, and during one flight with Joe I saw a waterspout right out the window. It appeared suddenly as a ghostlike column and disappeared behind a curtain of rain.

My yearly visits to Florida were eye-openers for me. The Cambridge–Boston area is not a fertile ground for mayhem of the tornadic type. In that area, a thunderstorm is characterized by the darkening of a hazy sky, followed by lightning, thunder, a cool gust of wind, and rain. Period. The viewing from Boston is poor at best. But in Coral Gables, Florida, thunderstorms as far away as a hundred miles or more could be viewed. Funnel clouds and waterspouts were almost routinely visible from the lab's fourth-story windows. I spent a fair amount of time perched on those window ledges, photographing them and other fascinating weather phenomena, and once my colleague Bob Burpee had to yank me back when, exhilarated, I teetered precariously close to the edge of the ledge while photographing a funnel cloud.

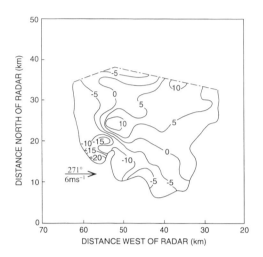

2.25 *The mesocyclone signature (roughly 53 km west and 22 km north of the radar) at 5 km above ground level as seen by the NSSL Doppler radar in Norman, Oklahoma. Storm-relative (storm motion indicated by arrow) Doppler velocities for the Union City, Oklahoma, storm of May 24, 1973 at 1536 CST. Note the approaching (negative) relative velocities in excess of 20 m/sec on the left and receding (positive) relative velocities in excess of 10 m/sec on the right, as viewed by the radar, which is off to the southeast of the storm. (Adapted from Lemon et al. 1978; courtesy of the American Meteorological Society. Copyright 1978.)*

The Union City Storm and Doppler Radar

In the early 1970s visual observations and Doppler-radar observations began to merge. A milestone in our understanding of tornadoes was reached on May 24, 1973, when the complete life history of a tornadic storm in Union City, Oklahoma, was documented in photographs, movies, and data from a pulsed Doppler radar at NSSL. By studying the evolving structure of the signatures of cyclonic rotation within the storm (Fig. 2.25), scientists at NSSL determined that the tornado was preceded by evidence of cyclonic rotation aloft by about forty minutes. This offered the first hope of improved short-term forecasts of tornadoes. There was also evidence of rotation in the tornado itself. This was called the *tornadic vortex signature,* or *TVS* (Fig. 2.26), which appeared aloft about twenty-five minutes before the tornado touched ground.

During these early days of Doppler-radar storm studies, all the radar wind-velocity fields being looked at were new, and radar data were not displayed in real time, as they are now. Instead, wind speeds were estimated (using the recent fast Fourier transform technique) long after the data had been collected.

The radar rotation signature was not obvious, although it had been detected before in other storms. Using radar with a wavelength of 5 centimeters, meteorologist Ralph Donaldson and his colleagues found a signature of the rotation, the *mesocyclone,* in a severe storm near Marblehead, Massachusetts, in the summer of 1968. They also noted a vault and an echo-free hole, the same features described earlier in the decade from non-Doppler data.

A mesocyclone is about two to five miles across and fits inside a severe thunderstorm. On the other hand, the vortex around an extratropical cyclone, which forms typically at the middle latitudes, is at least five hundred miles across. Most tornadoes are intense cyclones having diameters of 300 to 3,000 feet; hurricanes and typhoons are intense tropical cyclones with diameters of 150 to 250 miles across.

NSSL's Doppler radar first observed rotation in a storm on June 2, 1971, in central Oklahoma, and later in a squall line and a funnel cloud. The first tornadic supercell was documented on April 19, 1972, in Davis, Oklahoma, but the first observations by a pulsed Doppler radar while a tornado was actually in progress were made on August 9, 1972, in Chestnut Hill, Massachusetts, far from Tornado Alley. The radar was operated by the Air Force Cambridge Research Laboratory.

It wasn't until months after the Union City storm, in the fall of 1973, that investigators accidentally discovered the TVS. Rodger Brown, Don Burgess, and Les Lemon at NSSL were inspecting B-scans (tabulations of the Doppler velocity by range) for every azimuth and elevation angle. In trying, subjectively, to identify erroneous data points, they became suspicious of the data that displayed a large change in wind velocity over a small distance. But these so-called errors persisted in nearly the same location and at different heights; that is, the signature had temporal and spatial

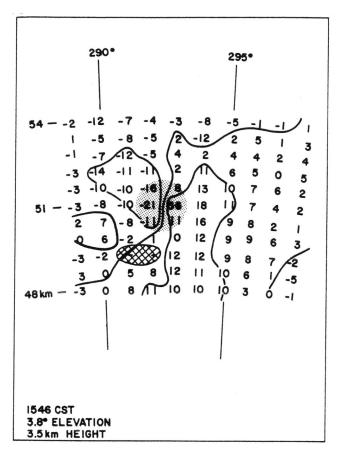

2.26 *The tornadic vortex signature (TVS) as seen by Doppler radar in the Union City, Oklahoma, storm of May 24, 1973. Mean Doppler velocities relative to the TVS are shown in m/sec. Velocities toward (away from) the radar are negative (positive). Azimuths and ranges are from the NSSL Doppler radar in Norman, Oklahoma. A weak-reflectivity, data-void area is hatched. (From Brown et al. 1978; courtesy of the American Meteorological Society. Copyright 1978.)*

continuity. Later experiments showed that the anomalous data correlated with the path the Union City tornado had taken. This radar signature eventually became known as the TVS.

Student Storm Chasing at the University of Oklahoma

A series of storm-intercept field programs continued at the University of Oklahoma through the 1970s, following the seminal Tornado Intercept Project. At MIT, I hadn't heard about the storm chasing underway in Oklahoma, but in my final year of graduate work I was fortunate to rub elbows with Ed Kessler, the founder and director of NSSL, who happened to be on leave at MIT. Kessler enthusiastically told stories of rotating storms in Oklahoma, bragged about the work meteorologist Joe Schaefer was doing on the dryline (where severe thunderstorms often form), and showered us with time-lapse radar film loops of the severe storms that occur day after day in Oklahoma. Kessler suggested that I come to Oklahoma to teach and to study severe weather. I liked Boston and wasn't thrilled with the idea of living in Oklahoma, but I couldn't pass up the opportunity to study tornadoes. So I arrived at the University of Oklahoma in the fall of 1976, fresh out of graduate school and burning with the desire to experience and study severe weather. Among the students I found there or heard about who eventually became my spiritual guides were Steve Tegtmeier, who, after witnessing so many incredible displays of severe weather, would become an evangelical minister; and John McGinley, Chuck Doswell, Gene Moore, Randy Zipzer, Mike Smith, and Al Moller, most of whom had been chasing storms as far back as the early and mid-seventies. The photographs Al Moller has taken over the years have wowed many an audience.

My first storm-chasing season began the following spring in a cooperative annual field experiment run by NSSL and the university's School of Meteorology. The university component was called the Severe Storm Intercept Project; I called it that rather than using the word *tornado* to prevent embarrassment in case we didn't see any tornadoes. That year there were two chase cars, station wagons leased from the university. John McGinley, an experienced chaser, was the leader of the first vehicle, named OU1. I was the leader of the second car, OU2—a strange choice, maybe, as I was almost completely inexperienced in chasing storms in the southern plains and dreaded sitting in a car for drives of more than an hour or two.

A primary objective of our expeditions was to try to provide ground verification for the 10-cm (S band) Doppler radar at NSSL. Suppose the radar indicated a mesocyclone signature or a tornadic vortex signature; was there in fact a tornado? If so, how was it related in time and space to the appearance of the Doppler vortex signatures? Was there large hail? Were there strong straight-line winds? Was there a tornado, hail, or strong straight-line wind not associated with Doppler signatures? Answers to such questions from our storm-chasing observations during this and following years became the basis for the warning procedures instituted by the National Weather Service in the early 1990s with the advent of the operational Doppler radar network in the United States. In 1977, there were only a few research Doppler radars in the country.

Doppler Dilemma

As the more seasoned leader, John McGinley with his OU1 crew got to roam anywhere necessary to find a severe storm. OU2 and I, however, were ordered to remain within the limited domain of NSSL's dual-Doppler network, which consisted of only a small portion of central Oklahoma. Remember that Doppler radar allows you to view the distribution of Doppler velocity, that is, *only* the wind component along the line of sight of the radar. With two Doppler radars, you view a storm from two different angles and thereby determine the wind field in the plane defined by two lines of sight of the radars and the *baseline* between the radars (the imaginary line connecting the two radars) (Fig. 2.27). To determine the actual three-dimensional wind (east-west, north-south, and local vertical), you need either another radar for a third viewing angle or a mathematical formula that relates the wind components to each other.

Actually, the situation is even more complex. Scatterers such as raindrops and hail have a terminal fall velocity; they accelerate as they fall, but eventually a drag force counteracts the force of gravity, and they begin to fall at a constant speed. The terminal fall velocity depends, among other things, on the shape and size of the raindrop or hail. Under typical atmospheric conditions, the terminal fall velocity of raindrops about a millimeter or so in diameter is 5 mph, while that of baseball-size hailstones can be 70 to 90 mph. Doppler radars determine these scatterers' components of motion, not the actual air motion. It turns out that with four Doppler radars, one can determine the three-dimensional wind field *and* the terminal fall velocity, but research Doppler radars are very expensive to build, operate, and maintain, so at that time we had to make do with only two radars: one at NSSL and the other twenty-five miles to the west at Cimarron Field (now Page Field) in Yukon, Oklahoma. These were 10-cm radars, which, less susceptible to attenuation by heavy precipitation than 3-cm radars (available elsewhere), were able to penetrate intense storms.

The problem of synthesizing the three-dimensional wind field from two Doppler radars was solved by using an empirical formula that relates the intensity of backscattering from precipitation to the terminal fall velocity, and an equation that expresses the conservation of mass, the equation of continuity. The former formula says that the stronger the signal at a given range, the larger the precipitation particles and the greater the terminal fall velocity. The latter equation says roughly that air squeezed together in one direction squirts out perpendicularly.

The two radars must be positioned so that the viewing angle of each is very different, to allow a dual-Doppler analysis. For example, the storm cannot lie on the baseline between the two radars. If it did, the second radar would add no new information in places where the wind is perpendicular to the baseline. Furthermore, the storm cannot be too far away, or the viewing angles to both radars will be nearly identical.

Generally, the best location to net a storm in the dual-Doppler area is along segments of circles lying on either side of the baseline. (If the wind is perpendicular to the baseline, the best locale for a storm is near the baseline bisector and away from the baseline; if the wind is parallel to the baseline, then the best place to find the storm is along the baseline.) Experience has dictated that the angle between the beams from each radar should be at least 40 degrees or greater. For the NSSL network, coverage was best over certain areas in central Oklahoma. OUi couldn't trek to the rich storm fields of the far western part of the state or the Texas Panhandle. Instead, we had to live off scraps of storms we snared in the local net.

The actual analysis of dual-Doppler data was quite time-consuming. Because the radar is pulsed, velocity measurements and range information can be ambiguous. The best velocity measurements are made when the pulses are transmitted in rapid succession. (Remember that velocity is measured by the Doppler shift of the backscattered signal. Suppose that

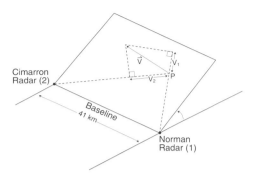

2.27 *The geometry of the NSSL dual-Doppler network. The baseline is formed by the straight line connecting the locations of the two radars (i.e., between the Norman radar and the Cimarron radar). The motion of precipitation (V) at point P in a tilted plane formed by the baseline and scans from each radar is resolved from the Doppler velocities from each radar (V_1 and V_2) as shown.*

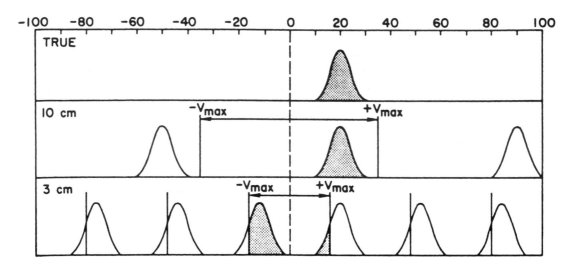

2.28 *An idealized illustration of velocity folding for 3- and 10-cm wavelength Doppler radars. For example, with the 10-cm wavelength radar, for the appropriate PRF, a velocity spectrum having a peak at 20 m/sec appears to also have a peak at -50 m/sec. For the 3-cm wavelength radar, a portion of the spectrum having velocities slightly in excess of 20 m/sec appears to have velocities slightly above -20 m/sec. The actual spectrum associated with a given mean Doppler velocity appears "folded" onto other portions of the spectrum having velocities of integer multiples of the Doppler spread (e.g., for a maximum unambiguous interval of $\pm V_{max}$ m/sec the Doppler spread is twice V_{max}). (From a technical memorandum by R. Brown in 1976. Courtesy of the National Severe Storms Laboratory of NOAA.)*

each cycle of the signal is sampled many times; then its frequency can be measured fairly accurately. If each cycle of the signal is sampled only a few times, then there could be oscillations of higher frequency that are missed; there is then ambiguity about the frequency of the signal.) The higher the transmission rate (the *pulse-repetition frequency*, or *PRF*), the higher the Doppler velocities that can be measured unambiguously. If the maximum unambiguous velocity were receding at 70 mph, then a velocity receding at 71 mph would be measured as approaching at 69 mph; the velocity measurement is said to be *folded* (Fig. 2.28) or *aliased* (known by another value). We can understand better how velocity folding works by thinking about a clock. One hour past twelve o'clock (for civilians at least) is one o'clock, not thirteen o'clock (as a military example, one hour and one minute past 2300 hours is 0001 hours not 2401 hours). So, for example, if the radar measured a wind velocity approaching at 50 mph, it could indeed be approaching at 50 mph, or it could be receding at 90 mph.

The least ambiguous range measurements are made when the pulses are transmitted at infrequent intervals. To get the range of the target, you send out a pulse and wait for the backscattered pulse to return. Since the pulse travels at the speed of light, you can calculate the range from the delay between the the transmitted pulse and the arrival of the backscattered pulse. If pulses are sent out too frequently, you may not know whether a backscattered pulse is from the most recent one transmitted and scattered off a nearby target, or the next-to-last pulse scattered off one farther away (Fig. 2.29).

The inability to make completely unambiguous measurements of both velocity and range is sometimes referred to as the Doppler dilemma. If possible, one has to identify and correct the errors due to velocity folding and errors in range (*range folding*). Correcting folded data and discarding hopelessly erroneous data, euphemistically known as editing, is time-consuming and boring. Its only virtue is that it is like solving a puzzle. Automated techniques and signal-processing schemes to eliminate most folding problems were not developed until the 1980s.

NSSL's dual-Doppler network snared its first tornadic supercell on April 20, 1974, near Oklahoma City. Unfortunately, a tornado formed when the storm was too close to the baseline between the two radars. The network's first tornadic storms to yield useful analyses occurred a month and a half later, on June 8, in the central and eastern parts of the state. Not for three years would another tornadic storm form in the NSSL network with the radars operating.

My First Tornado: The Cylinder Hypothesis

My first storm season was very frustrating. With little exception, the activity occurred outside the dual-Doppler area. On May 2 we did observe, from close range, a supercell as it split in half in Verden, Oklahoma, only fifty miles from Norman. Although we doggedly followed the wall cloud of the southern member of the storm, it refused to produce a tornado for us. But since Doppler radar data were being collected while we were out in the field, my first graduate student, Chris Sohl, was able to analyze the data and detail the splitting process, which had recently been investigated in computer simulations.

During these early days of chasing, the public knew virtually nothing of our activities. People often spotted us driving erratically, stopping, starting, moving quickly, or creeping along. Complaints to the university were frequent, and we were often warned to be more discreet.

And during the second week of May there was an episode of severe weather that came to be known as the "Seven Days in May," much of which we missed. Tornadoes raked the Texas Panhandle for fully a week, while we in OU2 had to stay close to home base. Finally, on May 20, after much begging, I was given permission to venture beyond the imaginary boundary of dual-Doppler coverage. I just *had* to see a tornado.

Storms began in earnest early on that day, well before midday, our usual time of departure. One tornado had completely destroyed an NSSL mesonet site near Fort Cobb—a touch of poetic justice! We headed southwest, hoping to pick up a new storm. Visibility was poor, and much of our strategy depended on Doppler radar information from NSSL. We were doing the ground-based equivalent of instrument flying (or flying by IFR-instrument flight rules). John Weaver back at the NSSL was the lab "now-

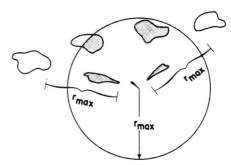

DOPPLER RADAR RANGE AMBIGUITY

2.29 *An idealized illustration of range folding. The radar is located at the center of the circle. Radar echoes on the radar display beyond the maximum unambiguous range (r_{max}), the unshaded echoes are "folded" back into the area within the maximum unambiguous range. Echoes actually within the maximum unambiguous range and the folded echoes both appear shaded. (From a technical memorandum by R. Brown in 1976. Courtesy of the National Severe Storms Laboratory of NOAA.)*

caster." A nowcaster has the responsibility of letting those on the road know the location of potentially tornadic storms, how they are moving, and where new storms are likely to form in the next five to ten minutes. In the Doppler shack at the lab, Don Burgess sat watching the display of the NSSL radar and tracking storms with mesocyclone signatures. Our contact with home base was by FM radio, using a repeater perched atop a television tower (destroyed by a tornado in 1998) in Oklahoma City. Radio was adequate for most of the dual-Doppler area, but outside it we used radiotelephone.

By late afternoon we were on the track of a tornadic storm, but we couldn't see it. According to the nowcaster, the tornado was heading for Tipton, in southwestern Oklahoma. From the storm's current motion and the time I needed to get to Tipton, I decided we might just get lucky and intercept it there. We turned west on a narrow paved road. It was hazy, and visibility was only a few miles. Suddenly, just to the southwest, the silhouette of a huge cylinder appeared, crossing the road ahead of us. It didn't look like any of the photographs of tornadoes I had ever seen, most of which looked like an elephant's trunk. Was this a real tornado, my first? I noted with great interest that the *cloud tags* associated with the cylinder—distinctive cloud fragments persistent enough to be tracked visually—were moving northeast at a speed faster than the cylinder itself. The cylinder was rotating! We stopped the car and I began to take pictures, furiously snapping the shutter. But I was still not convinced we were viewing a tornado; I thought maybe it was just a wall cloud.

Excitement—and inexperience—began to get the best of us. The member of our crew assigned to shoot 16-mm movies of the tornadoes we happened upon panicked and froze, unable to do anything. There would be no tornadic-debris movies to analyze. I somehow snapped the shutter of my still camera with no fear at all. As electric transmission wires flashed a pink-purple hue, I soon ran out of film, and in my haste I began to unload the exposed film before rewinding it. Fortunately, only several frames were lost as a result of my gaffe and the best images were not ruined. In a few minutes the cylinder disappeared behind rain curtains to the northwest. I still was not convinced the rotating cylinder was a tornado. I then instructed our driver, Dan Rusk, to continue driving so that we could determine if the cylinder really *was* a tornado. If we saw damage, it was a tornado; no damage, it wasn't. Just a quarter of a mile down the road atmospheric mischief was apparent: power lines were down, a house was missing its roof, and trees were delimbed. I had seen my first tornado! But my enthusiasm was tempered. The sight of real damage made me slightly queasy; I worried that someone might have been injured or killed, and that homes might have been lost.

I quickly realized I needed to contact our nowcaster, letting him know the exact location of the tornado and where it was heading so that he could notify the National Weather Service. But when I reached John Weaver, he told me, in essence, that he had more important fish to fry: A tornado was moving through metropolitan Oklahoma City, our own backyard. In fact, the parent storm had developed over Norman. But from where we were, we never had a chance of catching the Oklahoma City tornado. It all just goes to show that if you want to see severe weather, you should never let the weather gods know—they can be mischievous.

Tornadoes, however, were not the only prizes sought after. Large hailstones can indicate strong updrafts and have been collected in thunderstorms, especially in those being probed by Doppler radar. Nancy Knight, a cloud physicist, collaborated with NSSL with the objective of comparing the size of actual hailstones with those predicted by models. The success of her efforts in "Hail Chase" was limited in that the hailstorm had to be snared in the dual-Doppler network and she had to be on the spot to collect the hail—not an easy task.

3

Numerical Simulations Come of Age

When the wind produced in the cloud runs against another the result is similar to that produced when the wind is forced from a wide into a narrow place in a gateway or road. In such circumstances the first part of the stream is thrust aside by the resistance either of the narrow entrance or of the contrary wind and as a result forms a circular eddy of wind.

—*Aristotle,* Meteorologica (*on whirlwinds*)

THE DUAL-DOPPLER analyses of the tornadic storms of May 20, 1977, marked another milestone in severe-storm research. A partnership had begun between the numerical modelers, who simulate severe thunderstorms on a computer, and the observationalists, who analyze the data—both the visual data, such as photographs and movies, and quantitative data, such as Doppler-radar wind measurements.

We simulate storms so that we can "change" the environmental conditions in order to learn how those changes might affect a storm's behavior. Also, although dual-Doppler analyses gives us wind information, we do not have direct measurements of pressure, temperature, or moisture. In some cases, we can retrieve mathematically such thermodynamic variables from the Doppler wind-field analyses, but accuracy suffers if there is hail or a mixture of rain and hail. Furthermore, we do not have wind-field analyses at short intervals. It can take Doppler radar five minutes or longer just to scan the area of a single storm. Significant changes in the storm, such as the formation of a tornado, can take place in as little as a minute or two. With a computer model, we can diagnose the model's predictions of wind, pressure, temperature, and moisture and learn why what happened did happen.

Numerical simulations are based on a set of equations that express the physical laws responsible for storm behavior. The most important equation is Newton's Second Law of Motion. When applied to the atmosphere, this equation relates the force or acceleration of infinitesimally small collections of air molecules, which are assumed to act like a continuous blob of matter, called a parcel, to the forces acting on them. The equation of motion expresses the time rate of change of motion to all the forces acting on it. Because the rate of change of motion appears in the equation, the wind velocity is dependent on time.

Of all the numerous forces that can affect a parcel of air making its way through the atmosphere, the strongest are the *pressure-gradient force*—which is the result of horizontal and vertical differences in pressure—and buoyancy. The pressure-gradient force acts down the gradient, from relatively high to lower pressure. Other forces involved result from the earth's rotation and frictional drag. The former is tiny and is sometimes neglected in making numerical simulations of severe storms, but it is extremely important for larger-scale, longer-lasting features such as fronts, hurricanes, and extratropical cyclones.

In modeling, we relate the equation of motion expressed in a coordinate system that moves along with an air parcel to an equation expressed in a coordinate system fixed to the ground. When we do this, the acceleration depends on the product of wind components (or functions of them). This dependence is therefore nonlinear. Nonlinearity is responsible for the chaotic behavior of the atmosphere.

The nonlinear aspect of the equations presents a serious mathematical complication. In the numerical model, the effects of only slight changes in the wind field may be greatly amplified; consequently, forecasts made using the equations may be extremely sensitive to the initial distribution of wind, pressure, temperature, etc.

The equation of motion has one vertical and two horizontal components. The vertical contends with gravity. The horizontal components are usually taken to be the eastward and northward aspects. As there are many more than three variables involved in a thunderstorm—vertical and horizontal winds, pressure, temperature, water vapor, liquid water, ice—more equations must be used. The number of variables cannot be greater than the number of independent equations.

The equation of continuity relates the horizontal to the vertical components of the wind at any given time. The thermodynamic equation involves the conversion of heat into thermal energy and work, as manifested by changes in temperature, pressure, and density. Since air may contain water vapor, liquid water, and ice, equations describing the conversion of water substance from one form to another are also necessary.

The first attempts to simulate severe storms were limited a lot by the size and speed of the early computers. Modelers had to simplify the equations they used to build mathematical models in order to obtain results within a reasonable time. They ignored the effects of the earth's rotation and omitted sound waves. Today, although we don't believe sound waves affect the dynamics of thunderstorms, for mathematical convenience we include each sound wave oscillation, which typically occurs many thousands of times every second. (However, there is some evidence that sound waves emitted from storms, at pitches lower than we can hear, might offer clues of storm rotation.)

Some of the numerical models from the 1960s were one-dimensional and did not vary in time. Later models were two-dimensional, in part to conserve computer storage space and reduce the computer time needed to run simulations. The two-dimensional models produced a band of convection in which variations in atmospheric variables, such as wind and pressure, existed only with altitude and across the band. These models are fine for such two-dimensional phenomena as squall lines, but most severe thunderstorms are three-dimensional. In 1975 meteorologist Robert Schlesinger reported on his pioneering three-dimensional model of storms: He had obtained realistic results. At about the same time, Bob Wilhelmson and Joe Klemp developed a technique for solving equations with the inclusion of sound waves. Their three-dimensional cloud model became the standard for more than ten years. While inclusion of sound waves did not lend physical reality to their model, it did provide what they called "computational simplicity and flexibility." This meant that grids could be stretched to produce different spacings between points, according to the degree of resolution required. High resolution is needed in the vicinity of tornadoes and mesocyclones, for example, and near the ground, where the distribution of wind, pressure, and other fields exhibits high spatial variability.

Initially, the Klemp-Wilhelmson model appeared amazingly realistic in its simulation of the behavior of storms. But could it be run with a profile based on real wind measurements and produce a virtual storm that behaved like a real one? After the Seven Days in May, Klemp and Wilhelmson, along with meteorologist Peter Ray and his colleagues at NSSL, were determined to find out. Following several years of effort, they came up with a simulation of a storm that behaved similarly to the real thing—that resolved by dual-Doppler radar analysis.

Simulating Supercells

Keith Browning, a pioneering severe-storm researcher, proposed that there are two basic types of radar-echo cells associated with convective storms: the *ordinary cell* and the *supercell*. Both are the building blocks that can make up larger convective systems. A group of ordinary cells makes up a multicell storm complex. Ordinary cells are relatively short-lived, with lifespans of less than an hour; each cell can beget new ordinary cells, so the complex lasts for much longer than the individual lifetime of any cell. A supercell lasts much longer than any ordinary cell, and moves in a direction different from that of ordinary cells, usually to its right. It is nearly always associated with severe weather. Although usually isolated, supercells can also be arranged in a line, with gaps in between. In a relatively small area, such as the size of the state of Oklahoma, isolated supercells and multicells can coexist. What controls whether an ordinary cell or a supercell forms? The underlying assumption is that the two kinds of cells are controlled by different processes, and that each is unique to specific environmental conditions.

While Peter Ray, Joe Klemp, Bob Wilhelmson, and collaborators were attempting to simulate real storms in the late 1970s and early 1980s, meteorologists Morris Weisman and Joe Klemp began to explore the range of behavior that virtual storms might exhibit if idealized wind, temperature, and moisture fields were varied systematically, and if storms were triggered by an isolated buoyant bubble just above the ground. Observational evidence had shown that the different convective building blocks were related to the way in which the vertical wind shear varied with height. Vertical wind shear is a measure of how the wind speed and direction change with height.

Weisman and Klemp ran three-dimensional simulations on a supercomputer with fascinating results, showing a striking relationship between storm type and vertical wind shear, on the one hand, and the amount of buoyancy in a storm's updraft, on the other. The two proposed a classification of distinct types of storms based on the *bulk Richardson number* (bulk Ri). The bulk Ri is named after L. F. Richardson, an Englishman who early in the twentieth century attempted to forecast weather numerically by extrapolating a system of equations into the future to predict the distribution of wind and pressure. The original Richardson number described turbulent flows at a given height in the atmosphere. The bulk Ri, which is averaged over an air column instead of at a single level, is a measure of the relative effect of the buoyant energy in an updraft to the kinetic energy of the air flowing into the updraft near the cloud base. Using the Weisman-Klemp scheme, one could look at real data and attempt to forecast the type of storm that might occur.

A sense of the intensity of updrafts can be estimated by imagining that a bubble of low-level moist but unsaturated air is lifted first to the height at which water vapor condenses, then lifted even higher until it becomes buoyant. This virtual bubble is then allowed to move upward on its own, experiencing different rates of acceleration corresponding to the amount of buoyancy at each height. With height the vertical velocity becomes faster and faster until the buoyancy disappears and reverses in the lower stratosphere, where the updraft peters out. In real life the updraft in the troposphere is weaker because some cooler, unsaturated air is entrained into the bubble, and liquid water and ice load the bubble down. Actual updrafts in storms range in intensity from around 20 mph in weak storms to more than 100 mph in huge, fierce storms, which have considerable buoyancy in a deep column of the troposphere.

If the kinetic energy of air flowing into the updraft at cloud base, estimated from the vertical shear, is weak compared to buoyancy's contribution to the updraft, the bulk Ri is relatively high; a Byers-Braham ordinary cell forms, evolves through its life cycle, and dissipates (Fig. 3.1). Under certain conditions a gust front may trigger a sequence of new cells, each going through its life cycle and eventually dissipating (Fig. 3.2), so that a multicell complex forms. If the buoyancy is very high and the vertical shear

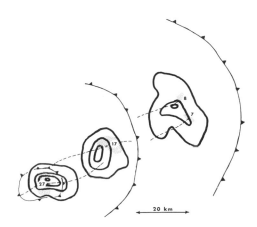

3.1 *Depiction of the life (in a computer simulation) of ordinary cells when the vertical wind shear over the lowest 5 km is weak compared to the strength of the updraft, which is modest. The storm structures are depicted at 40, 80, and 120 minutes into the simulation, which began with the initiation of the storm as a buoyant bubble. The storm positions are shown relative to the ground. Dashed lines represent updraft cell path. Low-level fields of rainwater, which can be related to radar reflectivity, are contoured. Surface gust fronts are marked by the meteorological symbol for a cold front, which is a line with "teeth" pointing from the cold air to the warm air. Regions of middle-level updraft exceeding 5 m/sec are shaded. Numbers at the updraft centers represent maximum vertical velocity (m/sec) at the time. (From Weisman and Klemp 1986, courtesy of the American Meteorological Society. Copyright 1986.)*

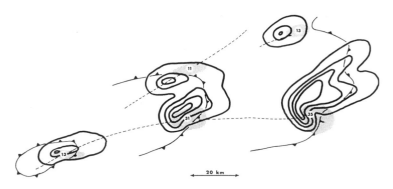

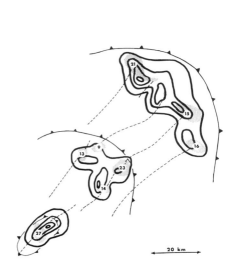

3.3 *As in Fig. 3.1, but for a case in which the vertical wind shear and updraft strength match. Suppose that north is to the top of the page. In this case, the storm splits into two parts: One (the northern one) becomes an ordinary-cell multicell storm, while the other (the southern one) becomes a supercell. (From Weisman and Klemp 1986, courtesy of the American Meteorological Society. Copyright 1986.)*

3.2 *As in Fig. 3.1, but when the vertical wind shear is higher and restricted only to the lowest 2.5 km of the troposphere. In this case the precipitation does not fall out into the updraft, as it does in Fig. 3.1, and the gust front triggers a succession of new cells. (From Weisman and Klemp 1986, courtesy of the American Meteorological Society. Copyright 1986.)*

is still weak, then large hail or strong straight-line winds at the surface or both could be produced.

If the vertical wind shear is increased for a given amount of buoyancy, the cell leans over (see Fig 1.10c). Precipitation no longer falls into the updraft, so the storm has a chance of living longer. The bulk Ri is smaller because the vertical shear is larger. If the vertical wind shear increases it may become so strong relative to the updraft that the latter starts to topple and is ripped apart from its roots below (see Fig. 2.9), the bulk Ri in this case is even smaller. There must be a match between the vertical shear and buoyancy if a storm is to be long-lived. If the bulk Ri falls within a certain range (about 10–15 to 45–50) a storm grown in such an environment could endure for a long time.

If both the vertical wind shear and updraft are large and matched, a supercell will form (Fig. 3.3). In this case the bulk Ri is relatively low and severe weather is likely.

How a Supercell Works

The matching of the effects of vertical shear and updraft intensity was explained through careful analysis of numerical simulations. In the early 1980s dynamicist Richard Rotunno worked with Joe Klemp in an attempt to understand why virtual storms behaved the way they do. They showed how the interaction between the storm updraft and the wind field outside of the storm controls the character of the storm. Earlier attempts to explain storm behavior had involved treating a storm as if it were a solid cylinder, sometimes rotating, with wind impinging on it in various ways. But storms are not solid objects, and the way wind interacts with them depends on pressure, which in turn is related to the three-dimensional wind field and buoyancy field.

We can understand what happens without looking at equations. First, consider the relationship between the wind field and the pressure field. If the atmosphere were hydrostatic, the pressure at any height would be given by the weight of the air above, which is determined by how its mass varies with height if it is not accelerating vertically. Mass at fixed pressure is determined by temperature. Horizontal motions are induced by pressure gradients that are consistent with variations in the weight of the air column, in horizontal gradients in temperature. But the atmosphere is not exactly hydrostatic, especially around a buoyant bubble that is rapidly accelerating upward. If it accelerates rapidly, as an elevator does when it begins to rise, then its weight increases, even though there is no increase in mass, just as your weight in the elevator does, even though your mass hasn't changed.

So part of the pressure is hydrostatic, a result of the mass (or temperature) field in an atmosphere that has negligible vertical accelerations, and part is due to dynamic effects, the result of accelerating and decelerating air currents around the bubble as it accelerates vertically. The lift of an airplane wing is due to the dynamic pressure field that is consistent with airflow around the wing. (The more advanced reader is referred to Appendix A for a more detailed explanation of dynamic pressure.)

Consider what happens when a buoyant bubble rises in an environment of vertical wind shear. Suppose that, for example, the low level winds are from the east and the upper-level winds are from the west; then the wind shear is westerly. This configuration of winds is accompanied (in the Northern Hemisphere) by cool air to the north and warm air to the south. Such a temperature gradient, from north to south, is the result of the mean pole-to-equator temperature gradient in most of the troposphere, and occurs because more radiation is absorbed at low latitudes than is returned to space, and more radiation is returned to space at high latitudes than is absorbed.

The movement of the buoyant bubble through the air changes its environment. And the behavior of the bubble changes, too, in a nonlinear way.

Imagine a paddle wheel with its axis perpendicular to the vertical shear vector (Fig. 3.4). It is easy to see that the paddle wheel will spin, as shown by the ribbons in the figure. A measure of rotation in a fluid such as the

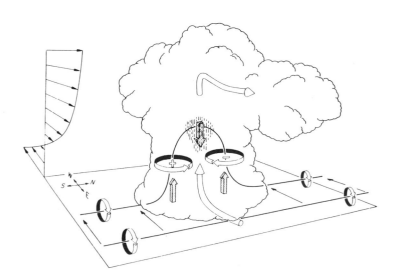

3.4 *An illustration of how horizontal vorticity associated with westerly vertical wind shear (the profile of wind as a function of height is shown in the rear left) is at least in part converted into a cyclonic-anticyclonic couplet (possibly manifest as counterrotating vortices) when an updraft grows in such an environment. The vortex lines indicate the axis about which there is vorticity; if you curl your right hand in the sense about which the air has vorticity (indicated by the circular arrow ribbons), then your thumb points in the direction of the vortex line. The updrafts induced by the areas of low pressure (the deviation from its hydrostatic value) associated with each member of the vortex couplet are indicated by the upward-pointing arrows. Direction of the cloud-relative wind flow is given by the cylindrical arrows. Initially the vortex line, which is aligned horizontally, is deformed as it interacts with the updraft of convective cell. (From Klemp 1987; adapted from Rotunno 1981.)*

atmosphere is called *vorticity* (from the Greek verb *vertere*, "to turn"). In this case vorticity is horizontal, because the axis of rotation is about the horizontal. In contrast, in most tornadoes and in hurricanes vorticity is vertical. An updraft carries air having horizontal vorticity. Along the edges of the updraft this rotating air starts to tilt, becoming more and more vertically oriented (Fig. 3.4). To the south of the updraft (or to the right of the vertical-shear vector), cyclonic vorticity develops; to the north (left of the vertical-shear vector), anticyclonic vorticity develops. As early as 1928 the German meteorologist Alfred Wegener, among others, postulated the importance of tilting a horizontal vortex onto the vertical in thunderstorms. But Wegener's vertical shear was apparently produced by the storms themselves in spewing out their anvils; Wegener named the source of vorticity the *mutterwirbl* (mother vortex).

The counterrotating vortices are strongest if the updraft and vertical shear are strong. Typically, the combined effect of updraft and vertical shear is strongest in the middle of the troposphere. What sort of pressure is associated with each vortex? If the air is to rotate about a vertical axis, there must be pressure-gradient forces turning the air constantly to the left in cyclonic flow and to the right in anticyclonic flow. Such forces imply that the pressure is relatively low in both cases. A minimum in pressure therefore develops at midlevels both to the right and left of the vertical-shear vector. It follows that an upward-directed pressure-gradient force is found to the sides of the updraft on the right and left of the vertical-shear vector (see Fig. 3.4). If air is forced up to its levels of condensation and

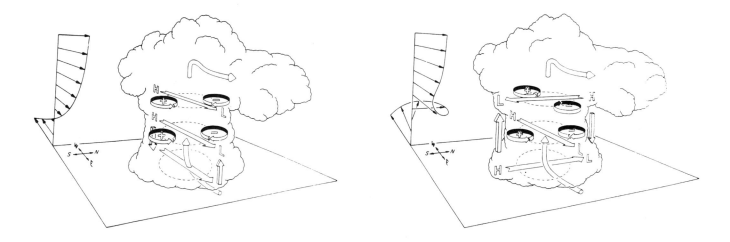

3.5   (*left*) *An illustration of how there is a net upward-directed (downward-directed) pressure gradient force on the downshear (upshear) side of the bubble depicted in Fig. B1. Solid arrows depict the orientation of the net vertical pressure-gradient forces. Flat arrows indicate the direction of the vertical shear vector in the environment at the given height. (Also shown are the preferred locations of cyclonic (+) and anticyclonic (-) vertical vorticity produced via tilting of horizontal vorticity, due to vertical shear, onto the vertical.) Direction of the cloud-relative wind flow is given by the cylindrical arrows. (From Klemp 1987; adapted from Rotunno and Klemp 1982.)*

3.6   (*right*) *An idealized illustration of the relationship between the pressure (the deviation from the hydrostatic value) field, the net vertical pressure-gradient forces, and the vertical wind profile when a buoyant bubble rises through an environment in which the vertical-shear vector veers with height. Otherwise, as in Fig. 3.5. (From Klemp 1987; adapted from Rotunno and Klemp 1982.)*

buoyancy, the storm will propagate to both sides, eventually splitting. Such behavior has been observed and simulated numerically with the vertical wind shear approximately unidirectional. Note that the average, or mean, wind in the troposphere is zero. In other words, the storm moves even though the mean wind is nearly calm. If either wind shear or updrafts—or both—were weak, the storm would not propagate much. Note that this scenario is independent of the effects of gust fronts. Real life is even more complicated.

The plot now thickens. In addition to the nonlinear tilting effect on the pressure field, there is also a linear effect. (The more technically inclined reader is referred to Appendix B for details.) The pressure must be relatively high on the upshear side and low on the downshear side of the buoyant bubble (Fig. 3.5). In unidirectional shear, new cell growth is stimulated on the downshear side and suppressed on the upshear side. If the vertical wind-shear vector *veers*, or rotates clockwise with height, new cell growth is favored to the right of the wind shear (Fig. 3.6); if it *backs*, or rotates counterclockwise with height, new cell growth is favored to the left of the wind shear. The former is case is most typical. Considering both the linear and nonlinear effects, we can understand how a storm might split, with the right-mover becoming dominant when the vertical wind-shear vector turns in a clockwise manner with height.

The region of midlevel cyclonic rotation or low-pressure area is the seed of the mesocyclone. An upward-directed pressure gradient underneath the area of rotation forces air up while it converges from all directions below (Fig. 3.7). The distance from the center of rotation below decreases and the rate of spin increases. This effect is analogous to the spin of a skater who brings in his or her arms closer to the body to spin more rapidly (owing to the conservation of angular momentum). But while the cyclone is forming, the updraft is acting on the vertical shear to form a cyclone on the right side of the updraft through tilting (Fig 3.8); a cyclone therefore appears to move to the right of the mean vertical-shear vector, by propagation, without becoming very intense. In order for the cyclone to intensify near the ground, something else must happen.

Let us suppose that upward motion occurs underneath a mesocyclone. Then air will converge under the mesocyclone and act to increase the rotation beneath it. The mesocyclone effectively lowers and the process is repeated as it descends. If a tornado formed aloft, it could descend almost to the ground by this process, which is called the *dynamic pipe effect*. It is a subject of some debate whether any tornadoes actually do form this way. Many form near the ground and build up, and others seem to appear at the ground and aloft at the same time.

Vertical shear is part of the Rosetta stone of the forecaster, who must decide whether to predict formation of a supercell based on observations in the real atmosphere. In the past decade, another related quantity has stirred up controversy among many scientists. This quantity is called *storm-relative environmental helicity*.

Helicity is a measure of the tendency of an updraft to rotate. It is high if the vertical wind shear is high and if a storm moves very differently from the mean wind. The deviant motion contributing to this condition is itself

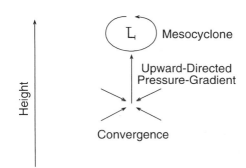

3.7 *An idealized illustration of the "dynamic pipe effect." The mesocyclone, which is strongest at midlevels in the atmosphere, is associated with a low-pressure area. A net upward-directed pressure-gradient force underneath the mesocyclone accelerates air upward. Convergence of air must drive the updraft; stretching increases the weaker circulation below. The increase in strength of the circulation below is associated with a drop in pressure below; a net upward-directed pressure gradient force below accelerates air upward, and so on.*

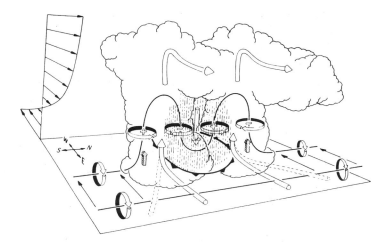

3.8 *As in Fig. 3.4, but as the storm is splitting. A downdraft forms in between the splitting updraft cells and tilts the vortex lines downward, producing two pairs of counterrotating vortices. The boundary of the cold air spreading out beneath the storm is indicated by the cold front symbol. (From Klemp 1987; adapted from Rotunno 1981.) (Courtesy of the American Meteorological Society. Copyright 1986.)*

related to the vertical shear of the wind (through the intensity of the midlevel vortices that are produced). And since some tornadoes may owe their existence to things that go on near the ground independent of the mesocyclone (such as gust fronts), storm-relative helicity may have little if anything to do with tornado formation and probably should not be used by forecasters as much as it currently is.

The central Great Plains of the United States is known colloquially as a tornado alley because this area is more likely to spawn supercell thunderstorms, prolific producers of tornadoes, than anywhere else in the world. The ingredients needed to produce a supercell are strong vertical wind shear, moderate to high values of buoyancy, moisture to feed the storm's buoyant updraft, and a mechanism for beginning the storm, such as heating at the ground or lifting stable air until buoyancy can be attained. If these conditions are present or if computer models indicate that they will occur, forecasters predict severe thunderstorms and tornadoes.

In the Great Plains, the jet stream, a band of very strong, high-level winds flowing generally from west to east in the troposphere, often prevails during the spring. Since winds at the surface are usually much weaker than they are aloft, the lower half of the troposphere is likely to have strong vertical wind shear. As air flows eastward across the Rocky Mountains, it sinks, is compressed, and warms. A warm core of air in the lee of the Rockies is associated with relatively low surface pressure. With this low pressure just east of the mountains, and higher pressure at the surface farther to the east, a pressure-gradient force causes air to move westward. After several hours, the effect of the earth's rotation is felt. Eventually, a state of balance (*geostrophic* balance) is almost achieved between the Coriolis force and the pressure-gradient force. The air then flows northward, bringing moisture and heat from the Gulf of Mexico to the central plains. Disturbances, each associated with relatively cool air aloft, propagate along the jet stream from west to east. With relatively cool air aloft and warm, moist air below, the environment is potentially unstable—if low-level air can become saturated and buoyant. The disturbances are associated with regions of rising motion. The rising motion cools off the air aloft and may deepen the layer of moist air at low levels. This results in a preconditioning of the air along air mass boundaries, such as fronts and the dryline, where even stronger upward motion can trigger storms. The disturbances also act to intensify the boundaries. In addition, enhanced vertical shear accompanies the disturbances.

Note that supercell behavior, in a dynamic sense, is the propagation of a rotating updraft to the right or left of the mean vertical-shear vector for much longer than the lifetime of an ordinary cell (which is longer than the time it takes a bubble of air to enter the base of the updraft in a storm and exit from its anvil). Until now, we have discussed this only in terms of the interaction of a buoyant bubble with its environmental vertical shear. For a more complete picture, we must also consider the effects of gust fronts produced as raindrops evaporate in unsaturated air.

The Effects of Evaporation of Rain

As the Thunderstorm Project showed, when rain evaporates in unsaturated air, a pool of relatively cool, dense air is produced. The dense air displaces the lighter ambient air ahead of the cold pool. An ordinary cell dissipates when its gust front propagates away from the cell's updraft, effectively cutting off the supply of warm air outside the storm from the updraft. In many instances new cells form from old ones along the edges of the gust fronts, and a multicell complex evolves. Under what conditions will cells keep on forming? Under what conditions will no new cells form?

In the mid-1980s, Richard Rotunno, Joe Klemp, and Morris Weisman, using numerical simulations, showed that without vertical shear below the top of the gust front (five thousand feet or so high), air that is forced up and over the evaporatively cooled air "leans over" so much that the air is lifted very slowly. But if there is vertical wind shear, and if it is pointed in the direction in which the gust front is moving, then air can be lifted straight up, enough perhaps to form a new convective cloud and cell.

In what is now called *RKW theory*, Rotunno, Klemp, and Weisman explained their results by considering the vorticity that is produced about a horizontal axis oriented along the gust front. When a shallow, cool (dense) air mass sits next to a warmer air mass, air along the boundary tends to spin upward in the warmer air mass and downward in the cooler one. This circulation is similar in nature to that set up in a room by a hot radiator. Eventually, the boundary leans toward the cooler air (Fig. 3.9, top) and the slope of air trajectories passing over the gust front is relatively shallow.

Now suppose that the vertical shear across the gust front is such that air is spun around it in the opposite direction to that induced thermally (Fig. 3.9, bottom). The air will flow higher up over the gust front and the slope of the boundary will be relatively steep. The deeper the lifting along a gust front, the more likely it is that air will be made buoyant and new cells will be triggered. We thus see that while supercells require strong vertical wind shear throughout a deep layer of the atmosphere for their longevity,

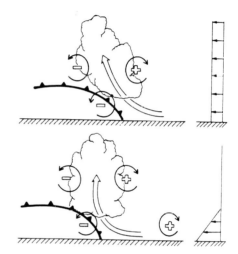

3.9 *An illustration of how the trajectories (thick arrow) of air over a gust front (cold front symbol) are more vertically oriented when the import of environmental horizontal vorticity (sense of rotation indicated by thin arrows) associated with vertical shear (indicated at the far right) counterbalances the production of horizontal vorticity associated with the temperature contrast across the gust front. A buoyant updraft is shown being triggered just above and to the rear of the edge of the gust front. Upper figure shows what happens in an environment having no vertical wind shear; lower figure shows what happens in an environment having low-level shear associated with horizontal vorticity of the opposite sign of that associated with the temperature contrast across the gust front. (Adapted from Rotunno et al. 1988; courtesy of the American Meteorological Society. Copyright 1988.)*

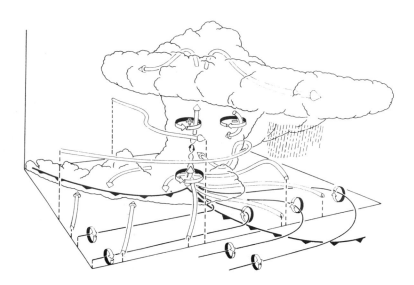

3.10 *An illustration of how horizontal vorticity produced along the forward-flank downdraft is tilted and spun up to intensify the low-level mesocyclone in a supercell. Three-dimensional schematic view of a numerically simulated supercell at a stage when the low-level rotation is intensifying. The storm is evolving in an environment of westerly vertical wind shear and is viewed from the southeast. Thin cylindrical arrows depict the storm-relative flow in and around the storm. The thin lines show the low-level vortex lines, with the sense of rotation indicated by the circular-ribbon arrows. The cold front symbol marks the boundary of the cold air beneath the storm. Air from the environment flows northward into the gust front depicted east of the updraft and then turns to the west and into the updraft. (From Klemp 1987.)*

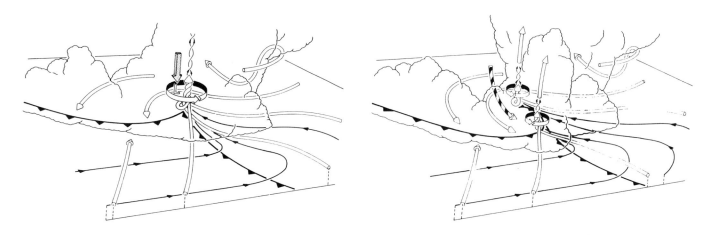

3.11 *An illustration of the origins of the rear-flank downdraft in a supercell from a downward-directed pressure-gradient force. Expanded three-dimensional perspective of the storm depicted in Fig. 3.10 at low levels. The figure to the left is valid at the time of Fig. 3.10; the figure to the right is valid approximately ten minutes later. Otherwise as in Fig. 3.10, except that the shaded arrow (left) represents the rotationally induced downward vertical pressure gradient force; the striped arrow (right) denotes the rear-flank downdraft, which is shifted slightly from the updraft. (From Klemp 1987.)*

3.12 *The dry slot and the decaying stage of a tornado in a supercell in northeastern Colorado, near Eckley, on June 8, 1994, at 5:20 P.M., viewed to the north from the NOAA P-3 aircraft.*

vertical wind shear near the ground is required for multicell longevity.

Evaporatively cooled air at low levels can play another important role in supercells. When wind shear is unidirectional, a mesocyclone will reach the ground only if there is a pool of cold air at low levels ahead of the mesocyclone. The mesocyclone in simulations forms at low levels when horizontal vorticity that is thermally induced along the *forward-flank downdraft* (Fig. 3.10) is acquired by air that is flowing toward the updraft. As the air enters the updraft, its vorticity is tilted so that it becomes vertical; it is then increased by convergence under the updraft. More recently, it has been suggested that horizontal vorticity may also be present near the ground outside the storm, having been generated somewhere else and then brought in. Depending on how it's oriented, it can enhance the formation of a mesocylone at the ground or work against it.

Eventually the low-level mesocyclone may become stronger than the mesocyclone aloft. The more intense the circulation, the lower the pressure, and so the pressure deficit at low levels becomes greater than aloft. A downward-directed pressure gradient develops, and this is responsible for driving a downdraft (Fig. 3.11). If the midlevel and lowlevel mesocyclones are not aligned vertically, air from the *rear-flank downdraft* wraps around the low-level mesocyclone and becomes what is called the *dry slot* (Fig. 3.12) as sinking air warms, reducing its humidity to the point at which clouds and precipitation disappear. The appearance of the dry slot is thought to be a sign that a tornado may form within minutes.* But not all dry slots are associated with tornadoes.

---

* Very recently it has been hypothesized that tilting of horizontal vorticity by the downdraft plays an important role in tornado formation.

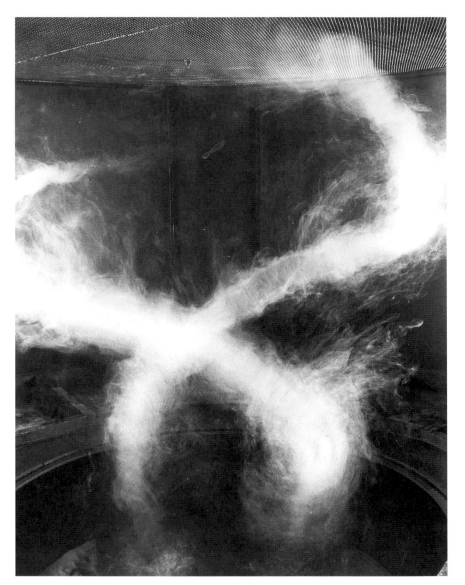

3.13 *A vortex chamber at Purdue University with multiple vortices visualized by dry ice. (Photograph courtesy of John Snow.)*

Simulations of Tornadoes

Tornadoes have been simulated in a number of different ways. In the 1950s and 1960s researchers simulated vortices using rotating tanks of water. Later, more realistic vortex chambers were built using air and fans. In the laboratory, only the vortex of a tornado is modeled, not its larger-scale parent. Ordinarily air in the chamber is sucked aloft by a fan, and rotation is imparted along the edges. The flow is made visible by dry ice or smoke from a generator. Smaller-scale cousins of lab vortex chambers are now commercially available, such as tornadoes in a bottle. You shake the bottles to generate rotation and a vortex that looks like a tornado forms in the froth.

Using a vortex chamber to study tornadoes is like studying one of your fingers when it is not connected to your hand. In order to make inferences about tornado behavior, we assume that tornado vortices are similar to the smaller-scale ones produced in a vortex chamber, even though the sizes and intensities of real tornadoes are different. However, a laboratory vortex is symmetrical, unlike most real tornadoes, around which air is pulled from different sources.

A significant challenge to researchers is measuring the wind, pressure, and temperature in the simulated vortex without disturbing the vortex itself. Pioneering quantitative work was first done by Neil Ward of NSSL in the 1960s. More recently, meteorologist John Snow and his colleagues have used various probes and a laser velocimeter to map successfully the wind and pressure distribution inside a large vortex chamber at Purdue University.

Vortex chambers do allow you to show how external parameters, such as the intensity of the exhaust (strength of an updraft) or the rotation rate of the outer portion of the chamber (intensity of the vorticity just before the tornado begins), affect the simulated vortex. Features seen in real tornadoes have been simulated in the chambers. For example, it was found that multiple vortices could be produced (Fig. 3.13) when the ratio of the wind speed about, or tangential to, the vortex exceeded the updraft speed by a certain critical amount. A measure of the ratio of the circulation about the vortex to the updraft is called the *swirl ratio*.

Another way to study tornadoes is to simulate a laboratory vortex on a computer—a technique that could be thought of as a simulation of a simulation. With a numerical simulation, one does not have to devise a scheme to measure the variables and risk disturbing the flow; all the variables are known. But the representation in the model of what happens where the vortex interacts with the ground needs to be done very carefully; the air flow there is turbulent and requires extremely high spatial and temporal resolution.

Consider what happens when a stationary vortex interacts with the ground. High above the ground, where its effects are small, the air flows around the vortex because the pressure-gradient force always acts to the left (Fig. 3.14a). The air accelerates just enough to the left so that it follows a circular path. But because the ground exerts a drag, air flow near the surface is slowed down, and instead of having a circular path, it spirals toward the center of the vortex (Fig. 3.14b). Think, for example, about what happens when you stir tea leaves in a cup. The leaves cluster near the center of the cup, converging toward the center of the axis of rotation of the tea—truly a tempest in a teacup.

We can also analyze the problem in the reference frame of the air itself. Imagine yourself hopping aboard the merry-go-round of the vortex. Here, well above the ground, the inward-directed pressure-gradient force is counterbalanced by an outward directed centrifugal force (see Fig. 3.14); recall from elementary physics that the centrifugal force is proportional to the square of the wind speed (and to the density of the air). Near the ground, the inward-directed pressure-gradient force overwhelms the outward-directed centrifugal force, which is markedly smaller because the

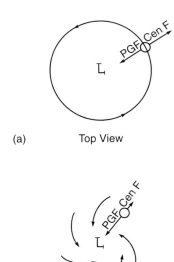

(a)   Top View

(b)   Top View

3.14 *Illustration of how frictionally induced convergence of air is induced when a tornado vortex rubs against the ground. (a) An example of how the pressure-gradient force (PGF) and centrifugal force (CenF) balance each other well above the ground in a tornado vortex. (b) The reduction in wind speed due to the drag of the ground reduces the centrifugal force so that it is less than the pressure-gradient force. The acceleration inward induced by the pressure-gradient force bends the trajectories of air flowing around the vortex with a radius of curvature that is less than that of the tornado vortex. The forces are indicated for a representative air parcel, which is denoted by a circle.*

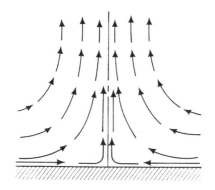

3.15 *An illustration of the idealized time-averaged wind field in the radial and vertical plane in a one-cell tornado, viewed in a vertical cross-section. (Compensating sinking motion away from the axis of rotation, the center vertical line, is not shown.) (From Davies-Jones 1986.)*

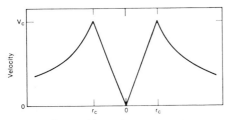

3.16 *The profile of the tangential component of the wind in a tornado vortex as a function of distance from the center in a combined Rankine vortex. The maximum wind velocities (V_c) are found at the "core" radius, r_c. (From Schwiesow 1981; courtesy of the American Meteorological Society. Copyright 1981.)*

wind speeds are weaker. Air spiraling in from all directions converges at the center and, since it cannot flow into the ground, it rises. The place where the air motion along the ground turns from horizontal to vertical is known as the *corner region* because this is where the air motion "turns the corner." The vortex thus takes on a one-cell appearance. Air rises in and near the center of the vortex (Fig. 3.15) and the compensating sinking motion takes place away from the center.

Consider now the vertical profile of the wind swirling around a tornado. Surface drag slows the wind down in the lowest several hundred feet. Just above the ground the air spins faster as it gets closer to the center, owing to conservation of angular momentum, but not indefinitely, because angular momentum is no longer conserved. It is not conserved because turbulent mixing becomes significant when the horizontal gradient of wind velocity increases, and mixing smooths out the wind field. Near the core of the tornado, which extends from its axis to the radius of maximum wind, the tangential winds behave as if the tornado were a rotating solid body. Imagine riding on a merry-go-round. If your horse is near the center, your tangential speed is slower than it is along the edge of the merry-go-round. As a result of this solid-body rotation, displacements into or out of the tornado are resisted. The resistance is due to imbalances between the centrifugal and horizontal pressure-gradient forces. The restoring forces that come into play if air is forced in or out support waves, which are sometimes seen traveling along the edge of condensation funnels. These waves make some tornadoes appear to dance around like snakes rising from a basket.

Outside the tornado core, the wind speeds drop off quickly at first, then more gradually with distance from the center of the vortex. The radial profile of tangential wind speed (which is a function of distance from the axis of rotation) is that of a *combined Rankine vortex* (Fig. 3.16). In the core there is solid-body rotation. Outside the core there is flow without vorticity. The vorticity due to curvature is canceled out by that due to the decrease in tangential wind with distance from the center. A combined Rankine vortex is a circular flow field having uniform vorticity embedded within an environment of no vorticity.

If the vortex spins more rapidly, its pressure deficit increases. Most of the intensification takes place just above the ground; most of the pressure falls therefore occur just above the ground. In fact, the pressure may fall so rapidly that a downward-directed pressure-gradient force develops and forces a downdraft in the middle. The abrupt change in the flow pattern in the core is called *vortex breakdown*. The downdraft may make it nearly all the way to the ground. In this case the tornado becomes two-celled. Air sinks relatively slowly far from the center of the vortex, rises closer to the center, and sinks at the center (Fig. 3.17). The diameter of the core of the two-celled tornado is wider than that of the one-celled tornado, and smaller satellite vortices may form.

If a tornado consists of cloud droplets and chunks of debris (which are much denser than air) lifted from the ground, the outward-directed centrifugal force, which is proportional to density, is greater than the inward-directed pressure-gradient force. The cloud droplets and debris are

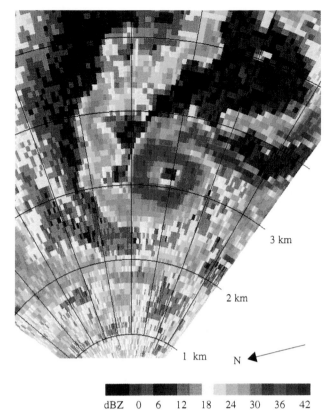

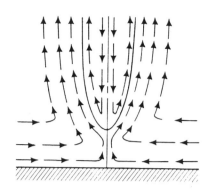

3.17 *Vortex breakdown and a two-celled circulation in a tornado, viewed in a vertical cross-section, as in Fig. 3.15. (From Davies-Jones 1986.)*

3 km

2 km

1 km   N

dBZ 0   6   12   18   24   30   36   42

3.18 *Weak-echo hole in a tornado in Dimmitt, Texas, at 7:04 P.M. on June 2, 1995, as seen by the Doppler on Wheels (DOW), a mobile, 3-cm wavelength radar. Radar reflectivity is shown in dBZ. The elevation angle was 2 degrees; this corresponds to a height of 110 m above ground level at the center of the tornado. (Courtesy of Josh Wurman, Jerry Straka, and Erik Rasmussen.)*

centrifuged outward, and a sheath of stuff flying around the tornado may form. The sheath should be detectable by radar. Radars close to tornadoes have in fact detected weak-echo holes (Fig. 3.18) that look like a miniature hurricane eye. And there have been tales of lucky observers, such as Will Keller, the Kansas farmer who looked up into a tornado and noted that it did in fact appear to be hollow.

4

Storm Chasing and Doppler Radar in Major Field Programs

Someday, a book may be written about storm chasers.
The longer that we do what others consider risky and unique,
the more superlative pictures that are taken, and the increasing
number of dramatic and humorous personal experiences—makes
this eventuality increasingly likely. It could be a very interesting
work—educational, exciting and philosophical; a thematically
unified piece with the customs and traditions of chasing—
all that describes and explains what we are and are about.
                                              —*David Hoadley,*
                                              Editor, Storm Track,
                                              Sept. 30, 1983

So FAR WE HAVE SEEN that advances in our understanding of severe storms have involved an intertwining of studies, some observational, some numerical. Our ability to observe severe-weather phenomena still surpasses our ability to simulate them. The challenge is to reveal undocumented phenomena and then to explain them. Never have numerical models simulated a hypothetical phenomenon that was later found in nature. And only recently did Brian Fiedler at the University of Oklahoma find evidence in numerical simulations that vertical velocities as high as the speed of sound could be present in suction vortices—a prediction, however, that has not yet been verified.

─◦◦─

My second season of storm chasing, the spring of 1978, was still focused on providing ground verification for Doppler-radar signatures. We operated in the Joint Doppler Operational Project (JDOP). Although the crop of storms that year yielded many beautiful photographs, it did not yield a proportionate number of tornadoes. We intercepted two nontornadic supercells late in April. One of them was especially interesting because at one point, when we were positioned next to a wall cloud the nowcaster reported that the storm was showing a TVS. Alas, we saw no tornado. Meteorologist Bob Davies-Jones, who was leading an NSSL team, was headed through the storm, toward our location, southeast of the wall cloud. A shy, reserved, Englishman, he had been doing research on tornado dynamics for a number of years, written or coauthored several authoritative review articles on tornadoes, and was the leader of NSSL's chase efforts. On hearing that he was heading right into the TVS, Bob replied to the nowcaster, who had cautioned him against going any farther, "I can't help it!"

On April 30, a tornadic storm hit El Reno, just west of Oklahoma City. Most of the intercept crews were southwest of Oklahoma City watching a storm that held promise of developing into a tornadic storm. The nowcaster, wanting to divert some of us from our passion of the moment, suggested that we get over to El Reno. Not meeting with an enthusiastic response, he implored at least one of us to seek out the storm. John McGinley in OU1 played the good soldier and left our storm to chase the other. As luck would have it, the El Reno storm produced a spectacular tornado; our storm did not. As they say in the stock market, bears win, bulls win, and pigs lose.

Toward the end of May, which usually marks the end of storm season in Oklahoma, and activity migrates northward to Kansas, eastern Colorado, and Nebraska, we became desperate and took off for west Texas. There, on May 26, we were rewarded big time when we tracked a storm from the New Mexico border, northeast across open country toward the head of the Palo Duro Canyon, just south of Amarillo. The canyon, unlike the fields of dirt to its west, looks like a smaller, lush, green version of the Grand Canyon. We stayed with the storm for more than five hours, watching it produce wall clouds and funnel clouds, but saw no tornadoes (see Figs. 2.14c and 2.17). After stopping for gas about 8 P.M., we finally caught a tornado and filmed it in the waning light. This marked our first successful solo chase, with minimal aid from the nowcaster. We felt as if we had truly earned our storm.

Exhilarated, we decided to continue the chase, since it was not yet pitch dark. But we were stopped on the highway and informed by a local spotter that a flash flood was imminent—good advice, for within minutes the entire highway was awash with runoff from the ten to fifteen inches of rain that had fallen on higher ground to the west. We spent the night in Plainview, intending to continue our chase by daylight. We were horrified by what we saw the next morning: Much of the road had been washed away, and debris was perched like ornaments on the upper branches of the trees (Fig. 4.1). Although we had gained confidence in our ability to chase severe storms and had documented some interesting events, we had not verified signatures for the Doppler radars back in Oklahoma.

4.1 *Flood damage south of Canyon, Texas, May 27, 1978, after a flood the day before. Note the absence of pavement in portions of the highway, the two cars washed away and deposited some distance from the road, and the freight train that never made it to its destination.*

The Discovery of LP Storms

Our storm chasing that season was not without some achievements. We also photographically catalogued cloud features much the same way nineteenth-century naturalists catalogued flora and fauna in newly explored lands. Some of our more interesting chases had nothing to do with tornadoes. On May 16, student crew member Cathy Kessinger shot a time-lapse movie of an isolated storm near Breckenridge, Texas. The storm's bell-shaped tower rotated cyclonically as it edged toward us (Fig. 4.2). A dark, opaque precipitation core was conspicuously absent, but baseball-sized hail fell. The storm evolved in an unusual way. The rotating base lowered as scud formed below, rose, and joined the cloud base above. The midsection of the storm became progressively narrower, almost tubelike, while the base grew wider. Viewed from the east, the storm was formed of layer upon layer of smooth bands. Apparently it had not dissipated in the traditional manner of the Byers-Braham cell, with a large, rain-filled downdraft destroying it; rather, rotation seemed to be playing a role. What we were seeing was the skeleton of a supercell, with all laid bare: rain playing the role of skin and clouds playing the role of bones. But the type of storm we documented had been noted a few years earlier. After our May 16 chase I heard about a similar storm that had produced a tornado in Tulsa, Oklahoma, in December 1975.

Don Burgess and Bob Davies-Jones suggested that little precipitation is associated with this type of storm. Since such storms have been noted near the dryline, the boundary between warm, moist air from the Gulf of

(a)

4.2 *Evolution of a low-precipitation (LP) supercell on May 16, 1978, northwest of Breckenridge, Texas, viewed from the east: (a) at 7:32 P.M. CDT; (b) at 7:44 P.M.; and (c) at 7:55 P.M.*

(b)

(c)

Mexico, and dry, warm air from the west, they named them "dryline storms." Yet tornadic supercells with lots of precipitation are also found near the dryline, so I suggested a more appropriate name—*low-precipitation (LP) storms*.

Meteorologists Al Moller, Chuck Doswell, and Ron Przybylinski coined the term *high-precipitation (HP) supercells* for those that produce so much precipitation that an opaque precipitation curtain forms to the rear of the wall cloud and flanking line (Fig. 4.3). We now believe that there is an entire spectrum of supercells ranging from those that produce hail but no rain at all at the ground—the LP storms—to those that produce copious amounts of rain—the HP storms. In an LP storm, the fall of large hailstones is spotty, allowing enough light through to produce a translucent precipitation core. Some supercells begin as LP storms but evolve into classic supercells, or HP storms. What accounts for the different ways precipitation develops, and why do LP storms tend to form along the dryline, not elsewhere, such as along fronts? Natural seeding of supercooled cloud droplets in growing clouds by ice crystals, may be involved. Since thunderstorm anvils produce ice crystals, it is thought that the wind structure aloft may play an important role.

4.3 *A high-precipitation (HP) supercell west of Gould, Oklahoma, at about 6:55 P.M. on April 18, 1992, viewed from the east, looking to the west. The area under the base to the left, which is opaque and filled with precipitation, is usually nearly precipitation-free and translucent in LP and "classic" supercells. A tail cloud is visible feeding into the wall cloud from the north.*

Another intriguing characteristic of LP storms is that little cloud-to-ground lightning activity seems to be associated with them, as is often found in rainy storms. Instead, rapid, staccato cloud-to-ground strokes of lightning are seen in the clear air under the anvil in the downshear direction from the storm's updraft. These lightning flashes might be called bolts from the gray (Fig. 4.4). Lightning is also seen inside the anvil. With so little lightning under the cloud base, an observer may never hear thunder. So an LP storm is an example of a severe thunderstorm producing little or no audible thunder at the ground.

Another intriguing find in our storm chasing in 1978 was that funnel clouds sometimes hang from high-based towering cumulus clouds (Fig. 4.5). These funnels, which are not associated with severe thunderstorms, appear to be associated with dissipating clouds having ragged bases. Circulation associated with them does not make it to the ground, and they don't become tornadoes, for reasons we do not yet understand. Even so, storm spotters occasionally report them to the Weather Service office, prompting unnecessary tornado warnings.

4.4   *Cloud-to-ground lightning outside of a thunderstorm cloud on September 4, 1980, south of Norman, Oklahoma. If this lightning flash had occurred during the day, it would have been a "bolt from the blue." This photograph demonstrates that you may be struck by lightning even if you are not directly under a thunderstorm.*

4.5   *High-based funnel clouds: (left) from a dissipating cumulus in the Texas Panhandle, east of Amarillo, on May 25, 1987, at approximately 4:17 P.M.; and (right) from the back side of a thunderstorm west of Salina, Kansas, on June 14, 1990, at 4:34 P.M.*

Gustnadoes

During JDOP we also noted short-lived, fifty- to hundred-foot wide dust whirls along the edges of gust fronts near the ground (Fig. 4.6). These features, which look like small tornadoes without condensation funnels, sometimes last for several minutes or more and seem quite menacing. Yet rarely are they strong enough to cause much damage. Storm chasers have coined the term *gustnado* to distinguish this phenomenon from its grown-up cousins.

SESAME

The following storm season, the spring of 1979, marked a major field experiment in the central United States, the Severe Environmental Storms and Mesoscale Experiment (SESAME). Besides the usual flotilla of chase vans and cars, there were also launches, from fixed sites, of *rawinsondes*, instrumented weather balloons that measure temperature and humidity and, through tracking, wind speed and direction, at levels up to fifty thousand feet or more. The rawinsonde is so named because a sounding (*sonde* is "sounding" in French) is determined from data that are radioed to the ground; wind information is also obtained by tracking the instrument packages. To improve the chances of snaring a storm, a third Doppler radar from NCAR was added to the two already in place at NSSL and at Page Field.

Severe storms were not frequent during SESAME, but when they did occur, they were spectacular. On April 10 the ingredients for severe weather were in place—an unusually intense disturbance in the wind, pressure, and temperature field aloft was propagating eastward from Utah, while a dryline was pushing rapidly in the same direction through west Texas. Expectations were high, and we were rewarded with major tornadoes in Texas and Oklahoma.

4.6 *A gustnado near the southwestern Nebraska–northeastern Colorado border, viewed from an NOAA aircraft on June 8, 1994 at 5:09 P.M.*

At that time weather data were still being churned out on rolls of paper from a teletype machine. Students huddled around the machine, eagerly awaiting the latest observations from key stations—which are usually located at airports (our ability to pinpoint features such as fronts and cyclones, in fact, depends on the spacing of cities with airports). That day, I watched intently as the machine clacked out rows of data. When I saw that the winds at Guadalupe Pass, a high-elevation station in southwest Texas, were gusting to over 70 mph from the southwest, I flew out the door with my crew and headed for Wichita Falls, in northwest Texas. The high winds belied the intensity of the disturbance aloft and the potential for strong lifting of air along the dryline, a necessary ingredient for setting off thunderstorms.

While on the road, we heard about a mesocyclone signature in a storm just north of the Red River, the border between Texas and Oklahoma. While during the 1978 storm season a relatively high percentage of storms with mesocyclone signatures had produced tornadoes within twenty to thirty minutes, a finding that would become the basis for the National Weather Service's warning procedures using operational Doppler radars a decade later, it was also becoming apparent to us that some storms have mesocyclone signatures only at higher elevations, and that they would not produce tornadoes. (We have since learned that mesocyclones in storms north of warm fronts often do not reach the ground and spawn tornadoes—probably owing to cool surface air, which weakens the updraft in the storm. With the often poor visibility north of warm fronts, it is difficult to document tornadoes even if they do occur.) Since we and the storm were both north of a warm front, now located near the river, I did not expect a tornado—certainly not one we could chase in the cloudy, gray, foggy, cool surroundings. But remembering the El Reno tornadic storm the previous year, the one that chasers had to have their arms twisted to go after, I diverted our team from our track and pursued the storm. However, we promptly decided that it had little tornado potential, broke away from it, and continued on to Wichita Falls.

When we broke through the warm front near the Red River, the sun came out and the temperature rose. At Wichita Falls we stopped to phone the nowcaster, who told us of storms to our southwest that were moving northeastward. We intercepted one of them almost immediately in Seymour, Texas, about fifty miles from Wichita Falls. A funnel cloud appeared from a wall cloud (Fig. 4.7), which was nearly overhead, and touched down as a large tornado to our west. It appeared to be white, illuminated by the sun at the rear of the storm, and soon was partially obscured by precipitation. We followed the tornado to the northeast for about fifteen minutes until it disappeared.

While our crew, the crews from NSSL, and freelance chasers gathered along the road to discuss the tornado, we noticed another tornado off in the distance to the northeast. It suddenly seemed absurd to be having a chat about a defunct tornado when new action was at hand. We raced back toward Wichita Falls but couldn't catch the new tornado. As we pulled into town, the destruction there from the tornado that had passed just minutes earlier was horribly evident. A stadium was heavily damaged, homes were destroyed, and trees and power lines were down. Hoping to catch up with the tornado as it headed toward the Red River, we backtracked to the highway and saw the magnificent back edge of that storm illuminated by the sun. It looked like an atomic bomb explosion. Then we heard that yet another tornado was hitting Lawton, Oklahoma, to our north. But there was no time to reach either one. We then decided to head home.

Twenty-five miles northeast of Wichita Falls, in southern Oklahoma, we found a field of debris—pieces of twisted sheet metal, canceled checks from a Wichita Falls bank, Wichita Falls newspapers, and pages from a grade-schooler's notebook. We later learned that canceled checks were found even farther away, in central and eastern Oklahoma. I collected

4.7 *Wall cloud viewed from the east, looking to the west, at Seymour, Texas, on April 10, 1979, at 4:40 P.M.; there is no well-defined tail cloud.*

some of the debris, which to this day resides in a box stored away in our building, as a souvenir of one of the most destructive tornadoes of the twentieth century. Forty-two people were killed and over seventeen hundred were injured in the tornado, which inflicted over $400 million in property damage and destroyed over three thousand homes. Sixteen people died in cars while attempting to flee the tornado. The damage path was fifty miles long; some F4 damage was noted.

Some tornadic thunderstorms act like bottles bearing messages that are dropped off in the ocean and retrieved somewhere on a far distant shore. More than fifteen years later, John Snow and his students at the University of Oklahoma conducted a systematic survey of all types of debris deposited some distance away from tornadoes and showed how paper items, which have small terminal fall speeds, could be transported far downstream to the left of the tornado track as a result of the wind structure both inside and in the environment of tornadic thunderstorms.

Although my crew had filmed the Seymour tornado, the NSSL chase crews and Gene Moore, a freelance chaser, filmed several higher-quality 16-mm movies of the event. Photogrammetric analysis of their films, which were taken from better vantage points than ours, where the tornado and its debris cloud were silhouetted against the sky, yielded estimates of wind speeds as high as 200 mph.

The next day, during a raging dust storm kicked up by high winds blowing over fields of loose soil, we drove back to Seymour to survey the damage. Although the tornado had struck mainly over open country, mesquite bushes had been ripped out of the ground and many dead cows lay in fields. One can only speculate that at least some of these cows had flown through the air. But the true intensity of the tornado could not be determined because few structures were hit, and those that were in fact struck were of unknown integrity.

That evening was the first night of Passover, and I had been invited to a seder that I nearly missed. My appearance late in the evening, after the ceremonies had taken place, was not appreciated by the hostess. Over the years my storm chasing would cause me to miss many social functions, including a party at my own home. Storm chasing is like gambling. You risk driving many miles, waiting idly for storms to form, and perhaps neglecting professional duties and personal and social obligations—all in the faint hope of catching a tornado. Most of the time you lose, but when you win, the memories of the losses quickly fade away.

Storm chasers also tend to neglect their health. Dinner is almost always late, delayed sometimes until 11 P.M. Although eating late is a custom in some European countries, our culinary experiences were far from Continental. We have had more than our fair share of fast-food hamburgers, pizza, and the like. Our worst meals are those from convenience stores. Miraculously, we rarely get ill. Some of our best meals, or at least the most enjoyable ones, have been at the Big Texan in Amarillo, Texas. There, as you eat, you can watch people swinging from the rafters and see card tricks performed right before your storm-glazed eyes. If you dare to eat their special seventy-two-ounce steak with all the trimmings within a specified time, the meal is free. Excellent barbecue eateries can also be found in Abilene, Texas, and in Ardmore, Ada, and Chickasha, Oklahoma.

The effects of the Wichita Falls tornado were still felt a year later, although indirectly. On April 2, 1980, the town was again threatened with destruction by a tornado approaching from the southwest. Some of the townspeople, whose memories of the tragedy the previous year were still vivid, panicked when a tornado warning was issued. A forty-one-year-old woman and her infant granddaughter sought refuge in a drainage canal, mindful that it was not safe to escape a tornado by speeding away from it. But, not aware of another tornado-related hazard, they were overtaken by a flash flood and drowned.

The next decent chase day came on May 2. We headed toward northwest Oklahoma, where a cold front was sagging southward and a weak dryline intersected the front. From experience we knew that the intersection of the dryline and a front is often a favored location of tornadic storms. When we reached western Oklahoma, a storm far to the north was poking up through the haze on the horizon. We barreled toward it.

Storm chasing often creates a roller-coaster ride of emotions. Our excitement on this occasion was intense. But when we reached the storm, it appeared that a gust front would choke off the supply of the warm, moist air necessary to sustain the storm. Our mood shifted to one of big disappointment. I called the nowcaster, John Weaver, to deliver our pessimistic report. His response was that a mesocyclone was off to our northeast. Once again our spirits soared—there was life left in this storm! But a wall of precipitation blocked our view of the action.

We then continued to the northeast via IFR (Instrument Flight Rules) (i.e. without visual guidance), with knowledge only of the location of the Doppler-radar signature—no significant cloud features were visible. Suddenly a huge wall cloud, under which dangled rotating curtains of rain, appeared to our southeast. We were on a collision path with the mesocyclone near the town of Orienta. We pulled east of the cool air under the storm into the warm, moist inflow ahead of it, positioning ourselves just north of the hook echo in the storm. The lighting was perfect. To our southwest, a multiple-vortex tornado appeared, silhouetted against a bright background (Fig. 4.8). Back to the west large hail was falling; if the tornado approached us, there was no retreating. One funnel cloud touched the ground and remained stationary, while another leapfrogged to its left, touched down, and rotated about the first one. Kerry Emanuel, a recent MIT graduate, calmly ran the movie camera. What I took to be birds flying around the tornado was actually debris—pieces of wood and fragments of plants and trees. Eventually the tornado consolidated into one condensation funnel and rampaged over largely open country off to the southeast, leaving us out of its path.

The footage of this multiple-vortex tornado, at the time among the best ever obtained, was later photogrammetrically analyzed. Wind speeds

4.8   *Multiple-vortex tornado west of Orienta, Oklahoma, on May 2, 1979, at 5:22 P.M., viewed from the northeast. Compare with the laboratory vortices seen in Fig. 3.13.*

were estimated to be as high as 170 mph. Only about six years earlier, multiple-vortex tornadoes were considered rare, even nonexistent. I recall a seminar given at MIT in the early 1970s by Ted Fujita, who pleaded for their existence to a skeptical audience of applied mathematicians and meteorologists.

The Orienta storm had passed inside the Doppler-radar network, making it possible to link our visual observations with the radar measurements. To the east another tornadic storm struck Fairview, Oklahoma; to the west of Orienta another refused to produce a tornado. By the luck of the draw, Bob Davies-Jones and his crew had to settle for the no-show.

Storm chasing can build up a lot of tension—much of it is released after the chase has ended, on the way home. During the early years of my chasing, from 1977 to 1979, there was considerable banter on the radio on the way home, and much of it took the form of imitations of one or two colleagues who had distinctive accents or speech patterns. Our radio could be monitored by local television stations and others who had scanners. Occasionally we crossed the line of decency and were reprimanded. The struggle between our need to let loose and radio etiquette was constant.

Some chase days are plain humdrum. Once, out in western Oklahoma, we stopped just east of the dryline to wait for storms to form. It was bright and sunny and not a cloud was in the sky. Inside the van some of the crew were asleep in their seats. Others lounged on the hood and roof of the van, reading magazines or books. A highway patrol car pulled up and an officer asked us suspicious-looking folk what we were doing. "Looking for tornadoes," we replied—and the patrolman broke out in loud laughter. It took some fast talking, but we managed to convince him that we were legitimate, after displaying all of our camera equipment and offering an impromptu lecture on severe thunderstorm formation.

During 1980, for the first time in a number of years, there was no joint NSSL–University of Oklahoma project. Much of the early funding for scientific storm chasing had come from the Atomic Energy Commission, which needed to know wind speeds of real tornadoes in order to design nuclear power plants that could withstand the force of tornadic winds— and then some. Apparently this was no longer an issue because new plants were not being built, and the funding ended.

Funding, in fact, was getting more and more difficult to obtain, and it became clear to me that we should use storm-intercept vehicles not only for collecting visual data, but also for making quantitative measurements in and near storms. Some within the meteorology community looked upon storm chasers as merely thrill seekers, engaged in a dangerous sport, and that other endeavors such as numerical simulations and the analysis of data from Doppler radar and aircraft were the serious science. Storm chasing needed a better image. Yet it was nearly half a decade before it got the respect it deserved.

TOTO and the Landspout

Probably the earliest *in situ* measurements of tornadolike vortices were made in waterspouts. During waterspout season in 1970 atmospheric physicist Chris Church and his crew flew over the Florida Keys towing a trailing-wire probe through waterspouts. Four years later, meteorology graduate student Verne Leverson, meteorologist Pete Sinclair, and Joe Golden, flying in an AT-6 acrobatic aircraft, used a gust probe to make measurements below the cloud base of mature waterspouts. They found that the pressure and temperature near the centers of waterspouts are lower and higher, respectively, than outside the waterspout—and they also found that air rises inside waterspouts and sinks outside. Tornado researchers took note of these observations.

In 1972 Bruce Morgan suggested that an armored vehicle could be used to penetrate a tornado to make measurements. This imaginative idea has never been put into practice, perhaps because it is perceived as crazy—a tank would probably inflict quite a bit of damage itself as it tumbles over the countryside, scurrying to get into a tornado. But at the American Meteorological Society's 1979 conference on severe local storms, Al Bedard, a scientist with NOAA in Boulder, Colorado, approached me with a novel idea. He and his coworkers had built hardened sensors designed to withstand local downslope windstorms, in which gusts can exceed 100 mph, and wanted to build a device that could be placed right in a tornado. It would take wind, pressure, temperature, and electric field measurements. Humidity measurements, which are more difficult to make, were ruled out because we could not find a suitable sensor. At the time, I was only vaguely aware of the earlier work on waterspouts.

With limited funds from NOAA and using spare parts, Bedard and his colleague Carl Ramzy built an ingenious four-hundred-pound device. A barrel-like base housed strip-chart recorders for the wind, pressure, and temperature sensors, which were mounted on a boom above (Fig. 4.9). A sensor for measuring corona discharge, a property of the electrical field, was also mounted on the boom. All measurements would be recorded in analog format. In light of our negligible budget, digital recording would have been too expensive. During the summer of 1980, with the aid of liquid stimulation at a cocktail party in Boulder, we named the device TOTO (Totable Tornado Observatory), after the dog that, along with its owner, Dorothy, was swept away in a tornado in *The Wizard of Oz*. TOTO was apparently the inspiration for a similar device named "Dorothy" in the 1996 movie *Twister*.

TOTO, which was mounted in the back of a government pickup truck, could be deployed in thirty seconds or less. In theory, we would get in the direct path of a tornado, roll TOTO down the truck's ramp, switch on the instrument package, and get the hell out of the way. One does not walk into a tornado holding up an anemometer and hope to survive! The tornado would, we reasoned, pass over TOTO, probably leaving it somewhat battered but (we hoped) still intact. We would then retrieve our mechanical canine and interpret the data traces. If the tornado did not change in intensity or size as it passed by TOTO, we could convert time to space and thereby determine the profile of wind speed and direction, pressure, and temperature across the vortex. We practiced deploying TOTO just east of the foothills of the Rockies in the vicinity of rather weak thundershowers during the summer of 1980, which was infamous for a searing heat wave in the southern plains of the United States.

The storm season of 1980 had not been eventful in Tornado Alley. For much of May, a blocking ridge diverted atmospheric disturbances elsewhere, and wind profiles were not conducive to tornadic-storm formation. It was not until the end of the month that the ridge broke down, briefly, out in west Texas, and a very strong flow aloft appeared. Severe storms began to brew. Erik Rasmussen, a graduate student at Texas Tech University, shot some impressive tornado movies in his own intercept vehicle, near Tulia and Lakeview, Texas. He also noted the cyclical tendency of some tornado episodes. We were nearing the birthplace of the tornadoes he filmed when our chase car, which had developed a hole in the floorboards, treated us to a geyserlike dousing as we drove into a flooded highway. That ended our chase—we had to limp home, missing the show.

4.9   *TOTO being tested in a wind tunnel at Texas A&M University on March 21, 1983.*

We first began to use TOTO on tornadic storms in the spring of 1981, one of the most active storm seasons for us ever. Our field experiment was not yet formally funded and our pilot experiment had been carried out on a shoestring budget. (The weather gods apparently took good notice of the lack of a formal, well-funded field program.) The first outbreak was on May 17, in central Oklahoma. Most of the chase vans from NSSL were focused on making electrical measurements. In our zeal to prove TOTO's worth, we headed north in pursuit of an early tornadic storm. But it was too far ahead to catch, so we retraced our steps, toward Oklahoma City, where we heard that another storm was moving through the northern suburbs. But heavy traffic and blinding rain delayed us, and by the time we got to the right spot, the tornado was long gone.

Next came our worst nightmare. Another storm was forming just southwest of Norman and was easily chased by those fortunate souls who had not jumped out early. While our radio blared graphic descriptions of the tornado, all we could do was listen. But another potentially tornadic storm was taking shape not far away from where we were, and we scrambled toward it. This time we made it and sent TOTO out under a wall cloud to make its first measurements. The storm failed to produce a tornado, but we had finally proved to ourselves that TOTO could be used in a storm.

Almost a week later, on May 22, Bob Davies-Jones at NSSL phoned just after lunch with the news that lightning was being detected in storms that had just popped up in western and southwestern Oklahoma, ahead of the dryline. He suggested that our group leave right then for parts west. I needed a good kick out the door, and Bob provided it.

Not far from overcast Norman, which was under a blustery south wind bearing abundant moisture from the Gulf of Mexico, one of our tires went flat, and we had to stop to change it. Perhaps the low pressure in the tire was a hint of what was to come. The weather gods had dropped a hurdle in our path while storms were building up to our west.

We successfully jumped the hurdle* and continued our caravan—the parent vehicle, a student's car, and TOTO in its pickup truck. At Cordell, about seventy-five miles west of Norman, we stopped to photograph a wall cloud to the northwest, viewed over a field of waving wheat. What followed was one of the most beautiful life histories of a tornado I have ever seen, timed perfectly for our arrival (Fig. 4.10). The truck carrying TOTO raced northward, but not fast enough to place it in the path of the tornado. Instead, in its first deployment with a tornado in sight, TOTO bore the brunt of strong winds along the gust front south of the tornado, but was not damaged and did not tip over.

It was a memorable day. We observed a total of nine tornadoes, more than my group has ever seen since. Although some of that number were only dust whirls with no condensation funnels, and some were too far

---

* Such hurdles are not uncommon. In May 1997, I was trapped in an elevator while trying to leave my office to go storm chasing. Fortunately I made it out in time to catch a storm.

(a)

4.10   *The life cycle of the Cordell, Oklahoma, tornado of May 22, 1981, viewed from the south. Airborne dust first became visible underneath the wall cloud at 5:18 P.M. (not shown). (a) Dust becoming airborne underneath the wall cloud at 5:20 P.M.; (b) a rotating dust column forming under the wall cloud and obscuring any existent condensation funnel at 5:22 P.M.; (c) a narrow condensation funnel has become well developed and visible as the rotating dust column disappeared at 5:26 P.M., perhaps because the tornado had moved away from a recently plowed field; (d) the tornado picking up a dust sheath again, which surrounds the condensation funnel at 5:26 P.M.; (e) the tornado "roping out" at 5:28 P.M.; (f) the dissipation of the tornado at 5:28 P.M. The entire lifetime of the tornado was only about ten minutes.*

(b)

(c)

(d)

(e)

(f)

away to attempt TOTO measurements, one was a monster. The last one of the day and the grand finale, this intense tornado moved through the town of Binger, forty miles west of Norman. It began as a multiple-vortex tornado and consolidated into a large cylindrical one (Fig. 4.11). Fortunately, it dissipated as it headed toward the outskirts of Oklahoma City. We tried to deploy TOTO right in the tornado but succeeded only in placing the device near it.

Back at NSSL, radar meteorologist Dusan Zrnic and his coworkers switched the Norman research pulsed-Doppler radar into a mode that allowed wind spectra to be computed in range bins every thousand feet. Ordinarily only the mean Doppler velocity and the radar reflectivity are recorded in a range bin. If a tornado appears within one or more of the bins and the Doppler wind speeds vary widely within the radar volumes (e.g., from 180 mph approaching to 180 mph receding), information concerning the *maximum* wind speeds is lost. In the high-PRF mode, spectra themselves are recorded. The Doppler-radar measurements indicated unaliased (unfolded) wind speeds as high as 200 mph in the Binger storm. Similar measurements had indicated weaker wind speeds in several other tornadic storms in previous years.

The chase ended at sunset, just west of Norman. Most of it had taken place along a single east-west stretch of highway. There was no long ride home and no early-morning arrival. It was a big success and our failures the previous week were forgotten.

The Cordell-Binger chase illustrates that you had better not arrive too early on the scene of developing convection along the dryline. It usually takes an hour or more for tornadoes to develop in supercells. If you drive right up to the place where a storm is forming, you run the risk of a tornado forming twenty-five to fifty miles off to the north or south in another storm and you might not be able to get to the right spot to intercept it. It is better to remain well ahead of the dryline, allowing time for the initial line of towering cumulus and developing thunderstorms to mature into individual supercells and then to move north or south to intercept the storm most likely to produce a tornado. The risk of staying too far back is that you may not be able to see the storms taking shape, and without good radar and satellite information you may not know where to head.

The 1981 storm season was important not only because it marked our first real use of TOTO, but also because of a serendipitous observation. On April 19 we set out, reluctantly, with a photographer and writer from *People* magazine who were eager to get a story about storm chasers. We went out mainly for the sake of the photographer and writer—we knew that the atmospheric conditions in nearby hunting grounds would probably not support the formation of tornadic supercells. But we also knew that in northwest Texas, conditions were more favorable. A storm with a mesocyclone was brewing there, so off we went—even though I felt that the odds of tracking down a tornado in the Texas storm were long. About midway to our destination I turned around to speak with one of our crew in the backseat. Astounded, I spotted a tornado out the *rear* window.

(a)

(b)

4.11 *The Binger, Oklahoma, "maxi-"tornado of May 22, 1981, viewed to the north-northwest, from a vantage point east of Binger. At 7:53 P.M. the tornado had begun as multiple vortices (not shown). (a) Dust was being kicked up under the wall cloud at 7:55 P.M.; (b) the mature stage of the tornado at 8:10 P.M., farther east of Binger than in (a). The tornado disappeared about 8:14 P.M.; the entire lifetime of the tornado was about twenty-two minutes.*

4.12 *A landspout along a developing line of thunderstorms in south-central Okla-homa, east of Waurika, on April 19, 1981, at 5:43 P.M. Note how narrow the condensa-tion funnel appears. In arid areas such as Colorado, landspouts often are not associated with enough moisture to produce a condensation funnel; instead, only a column of dust is visible.*

When I called Don Burgess, the operator of NSSL's Doppler radar, from our car phone, he informed me that where we were witnessing a tor-nado, there was no mesocyclone and hardly any storm! Instead, the radar showed what appeared to be only the beginning of a storm, and not much precipitation was reaching the ground.

Then yet another tornado formed behind us and, like the first, it hov-ered over open country. Both had very narrow condensation funnels (Fig. 4.12), reminding me of waterspouts I had seen in south Florida years ear-lier. The tornadoes looked fairly benign; at worst, I thought, they might unroof a doghouse. Subsequently I named small tornadoes in incipient thunderstorms *landspouts* because they look like waterspouts and appar-ently form under similar conditions. (Later I learned that someone had suggested the term "land waterspout" as early as 1927, but it hadn't caught on.) Landspouts contrast sharply with tornadoes in mature supercell storms, which are usually associated with mesocyclones. I was aware that Don Burgess, in correlating tornadoes with mesocyclone signatures in Oklahoma, recognized that some tornadoes have no such associated signa-ture. Now I had seen two of them myself. The guests from *People* magazine were disappointed that they didn't see a ripsnorting monster tornado fling-ing debris all over the place. At one point I was asked to pose, point to the tornado, and act frantic: I wasn't, and I didn't.

Landspouts are not the only nonsupercell tornadoes that have been documented. Gustnadoes, previously noted, are another. Funnel clouds are sometimes produced in convective storms that form beneath pools of rela-tively cold air aloft. These are called "cold-air funnels." Nonsupercell tor-nadoes have also been documented at the leading edge of a rainband along a cold front near Sacramento, California, of all places—California is not well known for its tornadoes.

For the 1982 storm season we obtained funding from the National Science Foundation (NSF) and NSSL to use TOTO in a cooperative effort with NSSL. Our parent vehicle, an old contraption once used by the University of Wyoming, we dubbed the Wyoming van.

On May 11 we initially headed west from Oklahoma City. But after hearing from Conrad Ziegler, the nowcaster at NSSL, that there was a tornadic storm in southwest Oklahoma near Altus, we took off in that direction; unfortunately no well-defined storm structure was to be seen. Soon it became clear that we would have to core-punch the storm—that is, drive blindly through heavy rain and hail to reach the area where there might be a tornado. It was risky, but to take a safer route might mean missing a tornado.

We now had the precise location and movement of the tornado itself and chose a route by which we could "thread the needle" through the storm, maximizing our chances of intercepting the tornado. If we took a safer route, the tornado might be too far off to see. The rain got heavier and heavier and small hail began to mix in with it. Then the hailstones got larger and the rain disappeared. Finally, there were only spurts of large hail, and visibility suddenly improved. Off to the southwest, several miles away, we could see a decent-sized tornado causing millions of dollars of damage to Altus Air Force Base. Here was a golden opportunity to drop TOTO directly in the tornado's path.

We did so. However, the weather gods realized what was going on and countered by forcing the tornado to dissipate before it reached TOTO. The tease got worse. Another tornado, forming by a process reminscent of Erik Rasmussen's cyclical tornadogenesis (birth of a tornado) model, appeared to our west. It quickly built into a spectacular multiple-vortex tornado (Fig. 4.13), causing considerable damage to homes in Friendship, a tiny rural town northeast of Altus. Our cameraman, Bill McCaul, grappled with NSSL's 16-mm movie camera, attempting to take what would have been one of the most spectacular tornado movies of all time. Alas, the camera jammed; the tornado, like Medusa, had turned it to stone.

Discouraged but not willing to give up, we headed north, hoping to drop off TOTO in the "tornado kennel." The Altus tornado had been moving northeast, as had the parent storm. So it was reasonable to assume that the Friendship tornado to our west would move that way, too. But it didn't—instead, it moved off to the northwest! We continued our chase for an hour, with the funnel in sight and the tornado oscillating between a single vortex and multiple vortices. The closer we got, the farther it retreated westward. I have never watched a tornado for so long, unable to get in its path.

Frustration continued on the next good chase day. On May 18 the Wyoming van broke down near Matador, Texas. Leaving the van overnight to be repaired, we piled into a private chaser's car for the long drive back to Norman. The next day we were able to retrieve our van, now in working order, and to resume our chase. We now had a choice: go toward the Texas

4.13 *Multiple-vortex tornado striking Friendship, Oklahoma, viewed to the west, on May 11, 1982, at 5:03 P.M. This tornado was three quarters of a mile wide. Five condensation funnels are visible; note how they all lean outward with height from the center about which they are rotating.*

Panhandle, where storms that were too far away to see were firing up, or chase an isolated storm that was clearly visible to our west. We decided to target the nearby storm.

The chaotic nature of storm chasing became apparent: Since we had to return to Matador, we were in a better position to chase the "wrong" storm. Had we not gone there, we would have headed straight toward the Texas Panhandle, where the northern storm produced an unusually wide and intense (and now legendary) tornado near Pampa. Tim Marshall and Jim Leonard, private storm chasers extraordinaire, still regale us with tales of "Pampa day." The Pampa tornado was around a mile across, and appeared from a curtain of rain and forced the chasers into strategic retreat. The isolated storm we went after produced virtually no rainfall, rotated, and did not produce a tornado. It was an LP storm.

The chase on Pampa day illustrates a fairly common problem encountered in pursuing storms along the dryline. When a number of storms pop up, which one to target? Often the southernmost is best, since interference from cold surface outflow from adjacent storms is nonexistent. A strategy that sometimes works is to go after the southernmost storm, wait until a tornado forms and goes through its life cycle, then drop south to pick up the next storm along the dryline. Unfortunately, sometimes (as on May 19) the southernmost storm is not necessarily a tornadic one.

On the last good chase day of the season, June 10, our target area was western Kansas. But intense activity, we heard, was forming in the northern Texas Panhandle, and we decided to rendezvous with NSSL crews heading there. Once again, as on May 11, we were forced to core-punch, since the storm was to our south. We caught up with Bob Davies-Jones's chase vehicle near Spearman, a town to which we would return a number of times in the following decade. Bob was able to position himself and his crew just ahead and to the east of the wall cloud while we played catch-up, just behind the wall cloud, but just ahead of large hail—a truly precarious position.

Our objective was now to get ahead of the wall cloud as fast as we could and deploy TOTO. A tornado was forming just to our south. Dust whirls were spinning toward us, on a possible collision course. Judging that we could *not* outpace the speed of the tornado, I ordered the chase van and the TOTO pickup truck to stop and deploy. Even if we were just behind the tornado, data collected there could be useful.

The decision was a good one. The tornado crossed the road just to our east, uprooting trees and snapping utility poles. Strong northerly winds blew down power lines, one of which struck the windshield of our van. We sat stock-still, touching nothing, while the power line sizzled, seared a line in the windshield, and bounced off the van. It was a close call. Residents of a nearby mobile home told us that they sought refuge in their underground storm shelter to escape the tornado and returned to find their home completely destroyed. Insulation and twisted sheet metal were strewn about. And so another storm season ended without a direct hit by a tornado on TOTO, the eager tornado dog.

The next year, in March 1983, we tested TOTO in a wind tunnel at Texas A&M University. Unless it was anchored down or widened, we discovered, TOTO could tip over at wind speeds as low as 100 mph and might not be able to withstand a direct hit by an intense tornado. Even so, we tried again later that spring to make measurements with TOTO unmodified. But 1983 was not a good year for tornadic storms in the southern plains. Only on one day, May 17, was there a large tornado, and it moved over mostly open country, inflicting little damage. Visibility was poor due to blowing dust, and we couldn't get TOTO anywhere near it.

By this time, press and media accounts of what we were trying to do had captured, apparently, the imagination of the public. One day out in rural north-central Oklahoma, just ahead of a storm, we set up TOTO on the front yard of a rural house. The owners must have recognized us and, guessing what we were doing, ran like hell for cover. Of course, the weather gods stepped in, and no tornado formed. Besides, it seemed as though tornadoes tended to avoid TOTO—perhaps the reaction of people seeing us arrive on their front lawn with TOTO should have been one of relief.

On another occasion, we were the ones who ran for cover. Fred Sanders, my mentor from MIT, was visiting and chasing with our team. Fred, who has sailed many times in the Newport-to-Bermuda yacht race, has a store of yarns involving the perils of big waves in tropical storms, nocturnal waterspouts, and other mayhem of the natural variety. I was eager to show him real severe weather storms such as he had never seen. We met a supercell in northwest Texas, but were badly positioned and were forced to core-punch to reach the wall cloud. We drove blindly through its heavy rain core, which eventually turned to large hail, and were lucky to escape without incident. When we broke out of the hail, there, just ahead of us, was a group of storm chasers, all watching us in disbelief!

After the 1983 season we gave up trying to use TOTO; it was just too difficult to get in the path of a tornado at the right time. NSSL tried the maneuver for a few more years, and on April 30, 1985, Lou Wicker and his crew almost succeeded near Ardmore, Oklahoma. Unfortunately, the tornado was just starting up when measurements were made, and the maximum wind speeds recorded were less than those taken south of the Cordell tornado back in 1981. In late 1986 TOTO was decommissioned, and today is a museum piece, on view at National Oceanic and Atmospheric Administration headquarters in Washington, D.C. Other approaches would have to be devised for obtaining direct measurements in tornadoes.

5

The Importance of Portability

The windy exhalation causes
thunder and lightning when it is
produced in small quantities, widely
dispersed, and at frequent
intervals, and when it spreads
quickly and is of extreme rarity.
— *Aristotle*
Meteorologica

D ROPPING AN instrument package
like TOTO in the path of a tornado was an ambitious effort with only mar-
ginal success. While TOTO was being used, Stirling Colgate, a physicist
interested in tornadoes, developed a much smaller, lighter-weight device
designed to be shot from a plane into tornadoes (the idea of using instrumented rockets to probe tornadoes can be traced to proposals first made by
a group of meteorologists at Purdue University in the late 1960s). Colgate's
probes, which contained miniature sensors to measure pressure, temperature, and the intensity of the electric field, had to be sent into the tornado
at almost the speed of sound to keep them from being diverted by the tornadic wind.

For three spring storm seasons, Colgate flew his Cessna, with rockets
mounted on the wing racks, attempting to penetrate tornadoes. He was
guided by information provided in real time from NSSL. But many of the
rockets, soaked after the plane flew through the heavy rains he frequently
encountered, misfired and missed their targets. And turbulent air was a
hazard; on one occasion strong turbulence in the path of air flowing into a
tornado forced him to make an emergency landing in a field. Eventually he
determined that the rockets were too lightweight and fragile to work prop-

erly, but the FAA did not allow heavier rockets because of the risk of one falling and injuring someone on the ground. In the end, Colgate abandoned his probes, just as we had TOTO.

Up, Up, and Away with Portable Rawinsondes

Just as we were concluding our experiments with TOTO, technology had advanced to the point at which portable balloon-lofted instruments were becoming commercially available. Before the early 1980s, instrumented packages attached to weather balloons (radiosondes) were released twice daily, at 1200 and 0000 UTC (Universal Coordinated Time), mainly at fixed sites separated by approximately 200 to 250 miles.

A radiosonde that is tracked by locating the direction from which its radio signal is strongest is called a *rawinsonde*. Wind direction and speed are estimated from the rate of change in location of the instrument package. Rawinsondes are only temporary residents in the atmosphere, since the balloons carrying them eventually burst due to differences in pressure as they rise. That point is usually in the lower stratosphere, above about fifty thousand feet. The instrument packages, which are very light and smaller than a shoebox, then plunge to the ground. Sometimes the packages are recovered and returned.

Using data from the National Weather Service rawinsonde network, a computer can draw a three-dimensional map of the temperature, moisture content, and wind field. The scales over which the data can be analyzed are limited by the spacing between the rawinsonde sites and the time interval between soundings. The vertical resolution of the rawinsonde, however, is adequate for our analyses, about thirty feet or less. Suppose you want to measure the vertical variation of temperature, moisture, and wind just outside or inside a severe thunderstorm. If you wait for the storm to arrive at a fixed rawinsonde, you would likely wait for a long time. Since severe thunderstorms are local events, it can be assumed that vertical variations in and near them differ from those farther away. Otherwise, why would there be a severe thunderstorm in one place and not in another?

In 1984 we purchased a portable radiosonde unit and set out to release, from our chase vehicle, balloons carrying instrument packages. The receiver was the size of a shoe box. Heavy tanks of helium for the balloons were carefully secured in the van to minimize the chances of an explosion (the helium was stored under very high pressure) if we had a collision. The signal from the sonde was an annoying, repetitive, short sequence of beeps, not unlike those heard over and over again in the popular old movie *Close Encounters of the Third Kind*. Greg Byrd, then a graduate student, was the guru of the unit, using it during the off season as well to release radiosondes into snow- and rainbands in winter storms.

The preparation for a balloon launch went as follows: About ten or fifteen minutes before the launch, we ran the unit with a series of instructions, followed by a paper calibration tape. Then we turned on the instrument package by inserting a battery and cut a length of string to attach the package to the uninflated balloon. Then we would inflate the balloon very carefully: If we didn't use enough helium, it would not rise rapidly enough (or at all) and the balloon could take over an hour to reach the upper part of the troposphere, a horror story for anyone with a type-A personality. If too much helium was used, the balloon could burst prematurely. We did not have the equipment to determine the exact amount of helium needed; we just tried to make a good guess. When actually launching the balloon, we had to see that it didn't make contact with anything, such as the chase vehicle, or else it could burst. Launching a balloon in a high wind was very challenging indeed.

On April 26 we tried our first launch into a tornadic thunderstorm near Guthrie, Oklahoma. A tornado was visible to the north, so we attempted our launch two to three miles south of the tornado, with a strong southerly wind at our backs. But we miscalculated. The balloon was underinflated, and we watched in disappointment as the radiosonde skimmed the top of a wheat field that separated us from the tornado. There was no second chance that day.

It was not until the next year, 1985, with support from NSSL and the National Science Foundation, that we successfully launched a series of radiosondes along the dryline, where we expected storms to break out, and just outside growing storms (Fig. 5.1). Some of these early launchings were documented on the PBS show *Nova,* in a special on tornadoes. Our main objectives were to determine the near environment of severe thunderstorms along the dryline, and why they formed or didn't form. In order to get wind information from our balloons, we resorted to old technology, the surveyor's optical theodolite, because it was less expensive. The theodolite is a small telescope attached to a tripod; the azimuth and elevation angle of the telescope are measured. In a former incarnation, our movie photographer, Bill McCaul, had received training as an architect and knew how to use surveying equipment. With a knack for painstaking detail, he tracked the balloons and recorded information from the theodolite every ten seconds or so. From a knowledge of the azimuth and elevation angles, and a good guess of the ascent rate of a balloon, the locations of the balloon can be determined.

5.1 *Glen Lesins (left) and Bill McCaul (right) launching a radiosonde on April 20, 1985, at 4:22* P.M., *ahead of a developing storm in western Oklahoma, west of Berlin.*

The most interesting early finding was that even when soundings just ahead of the dryline indicated explosively growing storms, the storm rarely materialized. It was difficult to find the smoking gun. We did show, as we had expected, that moisture variations in soundings were large, on the scale of five or ten miles. For example, one day we released two balloons, one along the dryline where towering cumulus clouds were becoming small storms, and another a short distance to the east, where there were no clouds at all. The soundings were markedly different. Key elements in determining where clouds and storms form are local variations in moisture and the physical processes responsible for them.

Other findings involved wall clouds and storm updrafts. In 1986, we succeeded in releasing a sonde into the updraft of a tornadic supercell for the first time, but we almost didn't. We were headed toward the Texas Panhandle, and as we approached the Texas border, we saw cumulus towers building near the dryline. Which of them would be most likely to grow into

5.2 *Two tornadoes for the price of one: an example of cyclical tornadogenesis. The first tornado near Canadian, Texas, on May 7, 1986, in its rope stage (left) and the second in its mature stage (right), viewed from the southeast, at 5:36 P.M.*

tornadic supercells? Don Burgess back at NSSL told us that the strongest storms were going up in the northern Texas Panhandle, near Miami. As the surface winds backed around to the east-southeast and escalated to a howl, we released a radiosonde to get an "environmental" sounding. And yet another storm was visible, coming up from the south. We were in a much better position for intercepting this new storm, since it would be moving toward us. But Don advised us that the storm to our north, which was obscured by haze and blowing dust, was of more immediate interest because it was more mature and looked more impressive on radar.

Our choice was to go north, and there in the haze a tornado appeared. It was the first of four (Fig. 5.2) we saw that day. After the second one (Fig. 5.3) we managed to release a radiosonde into the wall cloud; soon after yet another tornado formed, near the town of Canadian, and the instrument package got caught up in the storm's updraft, rising at the incredibly high rate of 110 mph. The updraft speed agreed reasonably well with estimates based on theory and our environmental and in-cloud soundings; the buoyancy of the updraft was estimated from the difference in temperature inside and outside the storm, which was as much as 18°F.

The fourth tornado dissipated in an unusual way. It assumed a ropelike appearance and the condensation funnel aloft vanished, leaving behind a phantom funnel touching the ground (Fig. 5.4). It appeared as if the tornado had broken away from its parent storm, but it is more likely that drier air had been entrained into the funnel aloft.

Meanwhile, when Bob Davies-Jones and his NSSL crew arrived at the border, where we had started our hunt, Don Burgess informed them that the new storm to the south of our storm had a mesocyclone signature. It made sense for them to go after this nearby storm, which on the Doppler

5.3   *The second tornado near Canadian, Texas, on May 7, 1986, during its mature stage about 5:41 P.M., viewed from the southeast.*

5.4   *The last tornado in its rope stage, north or northeast of Canadian, Texas, on May 7, 1986, at 7:38 P.M. The remnants of the tornado appear as a "phantom" funnel on the ground, with no visible connection aloft. Since the tornado is brightly illuminated by the sun, it appears to be a "white" tornado.*

radar looked as if it might be a good one, rather than to continue toward Canadian. But it fizzled out, while the Canadian storm continued to stir up tornadoes. Again, we were reminded that successfully intercepting tornadic storms is a chancy business.

The next day, buoyed by our success, we chased again. Our luck did not hold up. Virtually all storm chasers in the southern plains—those in the pursuit of science and those in the pursuit of thrills alone, headed to southwest Oklahoma, where the potential for tornadoes appeared to be the highest. As a member of two international folk-dancing groups, I was to attend in Edmond, a suburb of Oklahoma City, an exhibition by a famous group on a national tour, but I chose to go storm chasing instead. As luck would have it, southwest Oklahoma produced no tornadoes, but on the way home, the car radio blared that a tornado was moving through Edmond! My folk-dancing comrades from Norman, with no desire at all to see a tornadic storm, watched it form on their way to Edmond; in fact, they were forced to seek shelter from the tornado, which slightly delayed the dance exhibition. Their reactions ranged from amazement to sheer terror as they watched the funnel approach.

A week later we thought we had another opportunity to release a radiosonde into a tornadic supercell near Snyder, a small town in southwest Oklahoma. We stayed east of a wall cloud until it moved south of the road on which we were retreating. Rotating rain curtains appeared under the wall cloud, and to the south a tornado appeared. It was impossible to launch a balloon given the precipitation and short time for deployment. The Snyder storm turned into an HP supercell as it tracked eastward and dumped a lot of hail and rain. We were not aware that a visually spectacular LP storm, later to pass by Windthorst (south of Wichita Falls), Texas, was forming off to our south. This tornadic storm, documented by chasers Al Moller and Chuck Doswell, looked like a rotating cylinder.

Of course, it's not just the weather gods who are responsible for missed opportunities like this. Once, on the way toward a line of convective storms in western Oklahoma, we spotted a large tornado near Weatherford, off on the horizon. But almost out of gas, we were forced to stop in full view of the tornado. There, as we filled up, we watched with dismay as a storm perfect for launching radiosondes vanished. The moral of the story: Always be sure to top off the chase vehicle's gas tank.

Releasing radiosondes could pose the occasional hazard. Late in May of 1987 we were scrutinizing a thunderstorm that would soon produce a tornado near the town of Gruver, Texas. The time came to release our radiosonde. It rose promptly to the utility lines overhead and became entangled, the instrument package and its bright red balloon dangling in the wind like a hanged outlaw. But not all was lost. A couple of cowboys appeared, as if on cue, and lassoed the package—probably the first and only time a radiosonde has ever been captured Wild West style. Not having learned from this experience, we released the package again moments later, with the identical result. Clearly, the atmosphere was resisting observation.

A major shortcoming of our sounding technique was the old-fashioned optical theodolite we were using to get wind speed and direction. Not only

5.5   *University of Oklahoma graduate students releasing an NCAR M (mobile)-CLASS rawinsonde in the Texas Panhandle near Shamrock at 3:42 P.M. on May 31, 1992, as part of a special course.*

was the procedure painstaking, it didn't work if the balloon drifted inside or behind a cloud. Scientists at NCAR had recently developed a new technique, CLASS (Cross-chain LORAN Atmospheric Sounding System), for tracking weather balloons, one that made use of LORAN navigation aids. It required neither visual sighting of the balloon nor a moveable tracking antenna. Instead, the radiosonde could be located by comparing signals received from a number of LORAN stations. In 1987, Dave Rust, a scientist at NSSL, installed a CLASS unit in one of the chase vans, and began to collect rawinsonde soundings wherever and whenever desired. NCAR soon also made its own mobile CLASS van (Fig. 5.5). Our radiosonde unit was also mobile, but since it was relatively small compared to the CLASS and could be carried around, we referred to it as "portable."

The main problem with the CLASS soundings was that the LORAN receivers could not always lock onto enough signals to obtain accurate locations, especially with interference from the high electrical noise frequently found near severe thunderstorms. Rust's primary interest, however, was to obtain electric field measurements as a function of height in severe thunderstorms. His rawinsondes employed larger balloons to accommodate his electric field meters, in addition to the other standard radiosonde instrument packages. With a large sample of electrical soundings, meteorologists could also begin to answer questions concerning the buildup of electric fields strong enough to cause lightning.

The Sound and the Fury of Tornadoes

Beyond the conventional measurements of wind, temperature, pressure, and moisture taken in the vicinity of severe storms are those of a more exotic nature. Impressed by an audio recording of a tornado, Roy Arnold, a physicist, attempted from 1976 to 1981 to release audible-sound packages in the path of tornadoes. Known as Sound Chase, the project's goal was to determine whether wind in a tornado creates a unique sound wave, one that could be used as a signature for identifying tornadoes. Al Bedard, designer of TOTO, was also interested in sound waves generated by tornadoes, but in the infrasound region, at frequencies lower than those audible to humans. Very low frequency sounds, unlike audible sounds, can be detected up to thousands of miles from their source. Al has in fact recently found unique sound signatures in several tornadoes. Meteorologist Abdul Abdullah, in a paper published in *Monthly Weather Review* in 1966, theorized how vortices might trigger such sound waves.

The search for sound waves that could be generated by tornado vortices is ironic, given that some computer models that simulate severe thunderstorms intentionally exclude sound-wave generation to simplify and shorten computations and because it is not believed that sound waves play any role in tornado formation. In all my years of storm chasing, I have never heard the freight-train roar many say they have heard in a tornado. I did hear a low roar in the Binger tornado and am familiar with the swishing sounds of rain driven by near-tornadic winds against the chase van. I suspect that one has to be much closer, dangerously close, to a tornado to hear the fabled roar. If tornadoes do emit sound waves that can be objectively measured by a remote instrument, we would be able to estimate the wind speeds or even the nature of the wind field.

The LANL Portable Doppler Radar

My involvement with TOTO began when a scientist working in a different area, meteorological instrumentation, heard about our tornado studies and approached me at a conference. And my involvement with portable Doppler radars began after I had appeared in the 1985 *Nova* program on tornadoes and someone knowledgeable about instrumentation who had watched the show contacted me. A television appearance, like a presentation at a scientific conference, is an effective way to advertise one's work and to attract collaborators. But a disadvantage to having scientific work made so public is that eccentrics approach you with off-the-wall suggestions. Bombing tornadoes to weaken them is one (though German meteorologist Fritz Rossmann, apparently a serious scientist, did in fact suggest in 1960 that tornadoes could be disrupted by heating the upper part of the funnel with aluminum carried on a missile). And only recently, I was greeted at my office by a man carrying a small mouse safely housed in a box. The mouse, the man claimed, could forecast tornadoes. Then there are the letters that warn us of the dangers inherent in uncovering the mysteries of God's ways or claim that the writer has access to knowledge denied to most mortals. The annoyances public exposure can bring, however, are more than compensated for by the possibility of meeting new collaborators and the opportunity to spark an interest in science in young people who might otherwise find science dull.

5.6 *The Los Alamos National Laboratory portable Doppler radar and its components on June 1, 1990, in the Texas Panhandle.*

One day in 1986 after the *Nova* special had been aired, Tom Morton,[*] an engineer at Texas Instruments, called, suggesting that I might be interested in a portable, low-power, 3-cm-wavelength, continuous wave (CW) Doppler radar that physicist Wes Unruh and his group at Los Alamos National Laboratory were using for a project of a nonmeteorological nature.

So early in 1987 I went to see Unruh and his group. With support from NSSL from the DOPLIGHT (Doppler Lightning) project, an experiment whose objective was to verify radar signatures with ground observation, we began a pilot experiment to test a modified version of the radar. The system consisted of the radar instrument housed in a rectangular box, on which were mounted two parabolic dish antennas for transmitting and receiving, a tripod for the radar box, and a chest to protect recorders and batteries from the elements (Fig. 5.6). The radar, which was solid-state, transmitted only one watt of continuous power—but this was more than enough to detect precipitation three to six miles away. Since we didn't have to risk danger by getting closer than one or two miles from the path of a tornado, I refer to probing a tornado with a portable radar as "safe sensing."

Setting up the radar is quite simple and takes only a few minutes. Operating it properly, however, is an art. The radar is aimed by sighting the

---

[*] Morton was unaware that in 1985 Dusan Zrnic and his collaborators at NSSL had analyzed Doppler wind spectra from the Binger tornado, proposing that even higher-resolution wind data could be obtained with a portable radar that could be transported nearer to tornadoes. Zrnic, however, moved on to other meteorological problems and the proposal was not acted upon at NSSL.

target in a boresight, and the signal is monitored on stereo earphones. The Doppler wind spectrum is heard as a whooshing sound, with one channel containing information on receding velocities and the other on approaching velocities. The pitch of the sound is proportional to the wind speed. Point targets, such as birds or automobiles, are heard as pure tones, while distributed targets, such as raindrops and insects, sound noisier. If the signal is too strong, the sound is distorted, and the power must be turned down if accurate wind speeds are to be obtained. If only white noise (a hissing sound) is heard, the signal return is not useful. Whereas rawinsonde balloon tracking requires excellent vision, portable Doppler radar monitoring requires excellent hearing.

A video camera mounted near the boresite documents the cloud features the radar probes. The images are recorded on videotape, and the signals from each channel of the radar are recorded on the stereo audio channels of the tape. Doppler wind spectra are not available in real time; the audio signals are played back later in the laboratory, where they are digitized and processed by computer. The LANL system effectively re-creates what was possible with the one Smith and Holmes used back in 1958, except that it can discriminate between approaching and receding velocities and can be moved around easily out in the field. Nearly twenty years were needed to accomplish this. But, like the 1958 system, the LANL portable Doppler radar was unable, at least at first, to yield range information.

While we were fortunate to have a portable radar system, we had little luck with the weather. The spring storm seasons of 1987 and 1988 were extreme drought years in terms of supercell tornadoes. The month of April in both years was almost devoid of tornadoes in Oklahoma, although we did manage to test the LANL radar on one occasion in rain and with a distant funnel cloud. It was not until late May 1987 that we encountered a supercell tornado, and then it lasted only a few minutes and was not clearly visible. We were able, however, to collect data that showed wind speeds as high as 130 mph. Unfortunately, we did not have simultaneous videos of the tornado. On that day, the same day that cowboys lassoed our radiosonde from the utility wires, we were beset with even more problems. Returning from the chase we drove into a small Texas town—and a foot or two of water. Waterlogged, the van engine choked and died. We did get a ride to a motel from some good Samaritans, but it was in the rear section of a pickup truck, exposed to the driving rain. The next morning, as I was calling for a taxi to take us to Amarillo, lightning struck nearby and the phone jumped out of my hand and went dead. The lessons we learned on this chase were never to drive into deep water and not to use the telephone during a thunderstorm.

The 1988 storm season was not much better. Most of the month of May was spent under a ridge of high pressure, with the winds too weak for supercell development. Only on May 30 did supercell storms fire up in Texas. Then we tracked a supercell through a local dust storm into west Texas and got excellent data on a rotating wall cloud. It did not produce a tornado, however (Fig. 5.7). The data and visual documentation clearly showed that the Doppler spectrum shifted from the approaching side to

5.7   *Wes Unruh using his portable Doppler radar system near Morton, Texas, to probe a wall cloud to the northwest, on May 30, 1988, at 6:57 P.M.*

the receding side as we scanned left to right across the wall cloud. The exercise gave me confidence that the radar really was working and could deliver good data—if only we could probe a decent tornado.

There seems to be a law of conservation of storms. While storm production was very poor in 1987 and 1988 in the southern plains, eastern Colorado had a bumper crop. Storm season in that area is mostly during the late spring and early summer. In July 1987, landspouts formed only about fifteen miles away from a fixed-site Doppler radar near Denver. Meteorologist Roger Wakimoto, a participant in CINDE (Convection INitiation and Downburst Experiment), correlated photographs he had taken of one of the Colorado tornadoes with fine-scale single-Doppler analyses of the wind field. He found a tornado signature in the radar data and showed how the tornado had begun from near the ground up, in the absence of much precipitation. Ray Brady and Ed Szoke, of the government group PROFS (Program for Regional Observing and Forecasting Services) in Boulder, had demonstrated independently from a study of an earlier tornado near Denver that the parent storms of landspouts had no mesocyclones and did not form under the same high-shear conditions as supercells. Landspouts, or nonsupercell tornadoes, seemed to begin near the ground and work their way up to cloud base, in sharp contrast to the supercell tornadoes, which are often preceded by as long as thirty minutes by a mesocyclone aloft.

The Colorado observations, most of which were serendipitous, demonstrated that Oklahoma and Texas supercell tornadoes are not the only animals in the zoo. Nor are mesocyclones and supercells the only tornado producers. Yet without a precedent mesocyclone, how could the National Weather Service issue a tornado warning based on radar data? Many of us believed that this wasn't a matter for much concern because landspouts are relatively weak—or so we thought. There was a hint of chauvinism in the belief that the southern-plains supercell tornadoes were strong and Colorado landspouts were weak. Another puzzle was why landspouts are so common in Colorado, but apparently less so in Oklahoma.

After our second frustrating year in Oklahoma, the *coup de grace* came on June 15, 1988, when several landspouts occurred simultaneously in the Denver area. The tornadoes fell into fixed-site dual-Doppler coverage. Some of them were probed at a range of only twelve miles, the data reinforcing the notion that they derive their rotation not from a mesocyclone, as do supercell tornadoes, but from vortices that are present near the ground before the storms mature.

In spite of our disappointment with the poor storm crop in Oklahoma, and the apparent ease with which tornadoes sought out fixed-site networks in Colorado, we continued to improve the LANL radar. The next storm season just had to be better, we kept telling ourselves.

During the summer of 1988 Wes Unruh and his coworkers, using an ordinary video camera to trigger linear sweeps in frequency, modified the radar in an ingenious way to obtain both range information and wind spectra. The technique they used, called *FM-CW* (frequency-modulated CW), was based in part on work done by radar engineer Dick Strauch in the mid-1970s. When converted to audio, the sweeps sound like chirping birds: FM-CW radars are therefore said to be "chirped." They have the same so-called Doppler dilemma as pulsed Doppler radars: The latter send out occasional bursts of energy at one frequency, while the former send out energy continuously, but at periodically varied frequency.

With a standard video camera, the maximum unambiguous velocity of the LANL radar is 257 mph (both in the approaching and receding directions), while the maximum unambiguous range is only 3.1 miles. It follows that there had better not be highly reflective scatterers beyond that range or there will be range-folding contamination. The range resolution is only 256 feet but is good enough to resolve tornadoes five hundred to a thousand feet across. FM-CW signals are recorded on the videotape, and the audio channels, previously used to record the CW signals, are now used for voice documentation. The signal monitored on a video display looks somewhat like op art. The sense of motion is given by the way lines tilt in the vertical, the speed is given by the degree to which the lines slope, and the range is given by spacing between the lines.

In practice, the radar operator switches manually between the CW and the FM-CW mode. Visual documentation and scanning are done in the CW mode, and range information is obtained in the FM-CW mode. The operator must regulate the radar very carefully: Too much power and the display will appear unstable and unsynchronized; too little power and no bands at all will appear on the display. The data from the videotape are taken back to the laboratory, where they are digitized and processed by computer.

In the spring of 1989 we were on our own again, supported mainly by NSF; NSSL was, as in the previous year, only minimally involved. It was May 13 before we finally saw a spectacular tornado from beginning to end, near Hodges, in northwest Texas (Fig. 5.8). While most tornadoes move

5.8 *(left) University of Oklahoma graduate students (from left to right) Steve Hrebenach, Greg Martin, and Sam Contorno probing the beginning of the Hodges, Texas, tornado debris cloud on May 13, 1989, at 6:29 P.M. (right) debris cloud and condensation funnel.*

along with their parent storms, which in this case was to the northeast, the Hodges tornado appeared to be upended by a gust front, and the top of the funnel tilted over our heads into an almost horizontal position, while the bottom part of the funnel and debris cloud moved off to the southwest over mostly open country (Fig. 5.9). A young farm boy who had just been given a video camera captured a spectacular view of the tornado, looking up along the edge of the tornado funnel to the cloud base. This was one of many striking videos that have been shot over the years by people who just happened to be at the right place at the right time.

The image I took of my graduate students scanning the same tornado has been reproduced many times—on the cover of *Time* magazine (May 20, 1996), in the *Bulletin of the American Meteorological Society* (December 1989), and in the scientific literature (Fig. 5.10). It looks as if we are collecting data, but in reality the students are looking down at the ground, trying to figure out if a loose wire can explain why the radar is not operating properly. In fact, when the radar had been modified to operate in the FM-CW

5.9 *The Hodges, Texas, tornado of May 13, 1989, being upended by a gust front.*

5.10 *As in Fig. 5.8a, but a wide-angle view later on.*

mode, the bias of an operational amplifier had been set incorrectly, and no useful data were recorded. Nothing was accomplished scientifically, but I got some great photographs!

We recorded our first successful FM-CW data in a wall cloud in June 1989, near Floydada, in the high plains east of Lubbock, Texas. We had arrived too late to probe the tornado, which dissipated just before we arrived. In three years we still had not collected good data in a tornado clearly visible to us.

It was not until 1990 when, again with support mainly from NSF, we finally obtained a good data set on a tornado. The season began early—so early, in fact, that we were not prepared. On March 13, when many of us were away from the university on spring break, a tornado outbreak hit parts of Oklahoma and Kansas. Although many of the tornadoes were documented by those fortunate enough to have been around, no measurements were made near them because no instruments were available. One of our students filmed a large tornado just southwest of Norman, but the prize on that day, the most intense and photogenic, was a violent one that ripped through Hesston, Kansas, coincidentally the home of Wes Unruh's mother, who was not in Hesston at the time. Poor Wes had chased many miles with us and not seen any tornadoes; had he been visiting his mother on this day, he could have seen a monster tornado! His mother, almost 93 years old at the time, was anxious to get back to Hesston to see if her friends were unharmed. Disappointed that she had missed the excitement, she returned two or three days after the tornado and found that the retirement place where she lived had not been hit.

That season we split our efforts between releasing soundings and attempting to collect radar data. Despite a better crop of early storms, by the end of May we still had no tornado data. On May 31, storms began to spawn near the dryline in the northern Texas Panhandle, a location that had served us well over the years. Because I was considering two possible target areas and could not decide which to choose, we did not leave until about two-thirty that afternoon. It would take us at least four hours to get to the Panhandle. Yet we were optimistic—in late May and early June, there is good lighting for photography as late as 9 P.M., so we could still have at least two hours of chasing.

From western Oklahoma we phoned Larry Ruthi at the National Weather Service Forecasting Office in Norman and inquired about our target area, where an anvil rose on the horizon in the haze. Larry told us that a storm that had produced a tornado had apparently fallen apart. We then had to decide whether to cut our losses and go back home, or continue to the target area, still an hour and a half or more away, and hope for new storm development. Since it appeared that new anvil material was emanating from our target area, I decided to gamble. We forged ahead, but by the time we crossed the border into Texas, there no longer was an active cloud tower feeding the anvil from low levels. Had we been suckered by the weather gods once again?

But suddenly another storm sprang up to our west. It seemed as if none of the storms this day could progress very far eastward. Soon we saw staccato lightning flashes connecting the anvil to the ground just east of the cloud base, the latter which was part of a spectacular rotating cylinder; it appeared to be an LP storm. Then in the distance to the west, we saw a large tornado pendant from the north side of the cylinder (Fig. 5.11). The LP storm is truly one of the most visually magnificent works of fluid architecture in the sky. The best vantage point from which to view such a storm is usually ahead of the storm, looking back toward it. Such a view clearly shows the smooth striations around the cylindrical base.

In our haste, we sped along undulating, generally poor secondary roads, attempting to find a route other than the main road, which did not afford a direct path to the tornado. With the tornado just in sight ahead, we hit a bump, and the video camera I held in my lap rose up and smacked me in the nose. My nose was still bleeding, but I was unfazed, and we continued, albeit more slowly. I still managed to shoot video and take pho-

5.11 *Supercell near Spearman, Texas, on May 31, 1990, at about 7:10 P.M., viewed from the east: (left) with a 50-mm lens; (top) with a 28-mm lens. Note that the tornado appears underneath the northern side of the cloud tower. Large hail was falling to the right of the tornado in the area brightly illuminated by the sun.*

tographs. Eventually we positioned ourselves east of the tornado, which was near the town of Spearman, and collected CW data. Although the tornado soon vanished behind a shield of precipitation, we managed to measure wind speeds as high as 200 mph. The tornado was too far away, however, to collect FM-CW data. We mistakenly followed the storm eastward a bit, hoping that another tornado would form. One did, but in a new storm to the west. We should have been mindful that on this day storms dissipated when they moved eastward.

Meanwhile, the next tornado hit Spearman directly. Before we could get there the tornado was gone. Arriving in Spearman, we were greeted by a scared, wet dog and saw damage to trees and houses that had suffered a direct hit. The Spearman storm brought us closer to our goal of determining the Doppler wind spectrum as a function of range in a visually documented tornado; we did collect data on the next to last tornado, but from far away. Good CW data were obtained, but good FM-CW data were not.

The NOAA P-3 Airborne Doppler Radar

During the spring of 1991, our fifth season with the LANL portable Doppler radar, we were finally rewarded. NSSL was hosting a field experiment called COPS (Cooperative Oklahoma Profiler Studies). Despite the name, the COPS was not really about profilers, which are fixed-site Doppler radars with antennas angled so that clear-air vertical profiles of the wind can be obtained; profilers perhaps will someday replace rawinsondes as the source of wind information above the ground. COPS focused on the structure of the dryline, storm initiation, and storm evolution; profilers were simply part of the observing network.

One of the most innovative features of COPS was the use of an NOAA P-3 Orion aircraft, equipped with a 3-cm Doppler radar. This aircraft had been used for years to probe the structure of hurricanes, the primary advantage being that it could fly right up to the hurricane itself. There was no need to wait for the hurricane to come to the radar. Such a plane can follow a severe thunderstorm from its inception to its demise.

In the way the radar had been used previously, it had transmitted beams outward, normal to the direction of aircraft motion. To obtain an analysis of the wind field, the pilots flew flight legs that were perpendicular to each other. Thus, one could synthesize the horizontal and vertical components of the wind field using standard dual-Doppler techniques. But, since storms were being probed by only one radar from two different viewing angles, meteorologist Dave Jorgensen, who had been involved with the P-3 radar since its inception, named the technique pseudo-dual Doppler analysis. Unfortunately, the efficacy of the technique was severely limited by the time it takes to fly the two flight legs. The analysis was not always representative, especially if the feature being probed was evolving rapidly.

For the 1991 storm season, the scanning strategy was changed. Rotating antennas mounted in the tail of the aircraft produced beams that pointed both fore and aft (Fig. 5.12). Called FAST (Forward and Aft Scanning Technique), the radar maps forward- and backward-pointing cones.

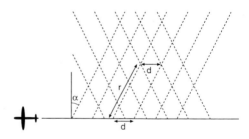

5.12 *The fore-aft scanning technique (FAST) scanning strategy of airborne Doppler radars: (top) θ is the angle swept about the aircraft and α is the angle (fore/aft) made by each sweep with respect to a plane normal to the flight track; (bottom) network of intersecting fore and aft beams; d is the spacing between consecutive fore/aft beams along the flight track; r is the range to a point of intersecting fore/aft beams.*

5.13 *Intense convection in the eyewall of Hurricane Diana off the Carolina coast on September 11, 1984, at 1:07 P.M., as viewed from an NOAA P-3 aircraft. Note the tilted bands, which lean from left to right with height and the "stadium effect."*

Data from intersecting beams are used to construct a pseudo-dual Doppler analysis. The main advantage of FAST is that only straight flight legs, not perpendicular ones, are required. At ranges under twelve miles the time difference between corresponding volumes scanned by the fore and aft beams is only a few minutes.

A recording network of PAM-II (Portable Automated Mesonet) surface observing sites from NCAR were also set up in the data-sparse Texas Panhandle and parts of Oklahoma to supplement the observing stations near airports. NSSL had M-CLASS vans in the field to release soundings. A new Doppler radar, the WSR-88D (Weather Surveillance Radar from 1988, Doppler), which replaced the old 1957 radar, collected data at Oklahoma City, and part of a new network of wind profilers from NOAA was up and running in Oklahoma, Kansas, and adjacent areas. The atmosphere was waiting to be observed.

My plan was to lead the chase with the LANL radar in our van and then, later in the field program, switch to the P-3 aircraft. I hadn't been on a research flight since 1984, when I had flown into the eye of Hurricane Diana off the North Carolina coast (Fig. 5.13) and into the eye of Hurricane Norbert off the west coast of Mexico. Bob Burpee had kindly arranged the hurricane flights. Chasing a storm at a speed of more than 200 mph seemed like a welcome change from the long, often boring rides on the ground. An additional benefit of chasing by aircraft is that not only can you watch a storm from a window, you can also see it depicted on radar and watch a running display in real time of wind, temperature, and moisture content.

The season began with a bang, even before the P-3 arrived. On April 11 weak storms formed along the dryline in western Oklahoma but failed to develop into carnivorous supercells. The next day we saw four tornadoes, one after another, in north-central Oklahoma. Unfortunately, the first one was too far away to probe with our portable radar. We reached the first one after we had decided to ignore the battle cries over our scanner from a notoriously unreliable source, who reported imminent tornadogenesis in a cell to our southwest. Instead we chose a cell to the north, in which, according to a more reliable source, there already was a tornado. It turned out that we made the correct move: The storm to our southwest did not produce a tornado. Gathering information in the field and making split-second decisions are crucial to successful storm chasing. In effect, we act like an intelligence unit and make decisions based upon our atmospheric informants.

On this particular day, one of our best out there, a local television crew was following us and documenting our tornadic odyssey. The next tornado formed to the northeast and arched itself across the road as it dissipated. We weren't able to catch it. Then to our west, a third tornado formed, in a perfect location for collecting data. Satellite vortices briefly rotated around it. Excellent CW data and visual documentation were recorded for about seven minutes, which seemed like an amazingly long time (Figs. 5.14 and 5.15). It was our best dataset to date. Doug Speheger, one of our crew members, uttered the words, "Oh, what a classic!" The images and his exclamation, which aptly described the tornado, by now have been reproduced on commercial videos and shown many times on television.

The next tornado looked larger (Fig. 5.16) but was partially encased in precipitation. It was not easy to intercept, but not for meteorological reasons: Local police had blocked the road leading to it. After briefly pleading our case, and convincing them that this was a field experiment, we were allowed to enter the domain of the hallowed vortex. But by then we were in dire need of gas. Stopping to refuel meant that the tornadic storm pushed on toward Kansas without us. The magic tornado door had shut.

Just two weeks later, on April 26, the tornado door opened again, with a vengeance. For more than a week a computer model at the National Meteorological Center in Maryland had been forecasting a major atmospheric disturbance in the southern plains. The P-3 had arrived, but, owing to an engine problem, it never took off. Those of us chasing in the land war were given a better chance.

The afternoon got off to a slow start. First, I shot some video of our students getting excited and making chase plans. Then we took off, with a TV crew from Tulsa following us. Crews often followed us, hoping to obtain spectacular video footage of tornadoes. Eventually we phoned NSSL for information but heard nothing encouraging. Why hadn't tornadic storms formed yet? I had expected an early show. Atmospheric conditions

5.14 *University of Oklahoma graduate students Doug Speheger (left), Jim LaDue (top), and Herb Stein (bottom right), probing the tornado north of Enid, Oklahoma, on April 12, 1991, with the portable Doppler radar.*

(a)

(b)

(c)

5.15 *The third tornado, north of Enid, Oklahoma, which we viewed on April 12, 1991, and the first one of the day we probed with the portable Doppler radar: (a) rotating debris cloud underneath the rotating wall cloud, to the west at about 4:55 P.M.; (b) "classic" tornado at about 5:00 P.M.; (c) wider-looking "classic" tornado shortly thereafter.*

5.16   *The fourth tornado we viewed on April 12, 1991, at approximately 5:35 P.M., east of Pondcreek, Okla-homa.*

were potentially explosive. The National Severe Storms Forecast Center in Kansas City had issued tornado watches for parts of Kansas and Oklahoma. Everyone was on high alert, waiting for the fuse to be ignited.

We continued north, where our chances of catching tornadoes were greater. Soon we found ourselves crossing the Kansas border amid towering cumulus clouds, the seeds of thunderstorms. The towers tried and tried but did not succeed in developing into full-blown storms. We were getting discouraged.

At 4:15 P.M. we heard that the storms were forming back in northern Oklahoma. We turned around and headed back to Oklahoma—and missed a rendezvous with an infamous tornado that plowed through parts of Wichita, Kansas. Videos from this storm, seen again and again on TV and commercially available tapes, show debris flying through the air over McConnell Air Force Base, and another tornado near El Dorado with people seeking shelter under an overpass as the funnel passed overhead. A trailer park in Andover, Kansas, was demolished and seventeen people lost their lives. More than $150 million in damage was done. But we had rolled our dice on the Oklahoma storms.

At 4:53 we caught up with a storm in north-central Oklahoma. A funnel cloud formed but soon dissipated; we had witnessed an aborted take-off. We followed the storm eastward. Then off to our south a new storm formed. First two thin, short-lived funnel clouds appeared, reminiscent of landspouts. Then at 5:47 another funnel cloud appeared, now to our east.

As we rushed to catch up to it, I called the National Weather Service about it. We then heard a severe-thunderstorm warning, again to our south. We changed course, and once more we missed a rendezvous with another powerful tornado, this one which formed south of Winfield, Kansas. So far we had sacrificed two tornadic storms in the hope of collecting data in another.

Our first glimpse of the new target storm to our south was near Billings, Oklahoma. As in any supercell storm, when you approach the storm from the north, you are likely to end up punching the core. We could hear the tornado sirens in Billings wailing, and it was ominously dark to the south. Hailstones as large as baseballs bombarded the van. I didn't like what I saw. If we went south, we might drive straight into a tornado. I couldn't locate the tornado precisely, and without a convenient road south, I could not thread the needle. I could only gamble, but I didn't like the odds. We retreated and drove east from Billings, trying to get ahead of the storm's path. Eventually the precipitation thinned and we could see a tornado off to the southwest; it was fortunate we had decided not to continue south.

We crossed the interstate and the tornado became more clearly visible. It was huge, and heading in our direction. I hoped that motorists on the road were awake and would not continue into the path of the oncoming tornado. After heading east a short distance, we turned south on Highway 77. The question now was how much farther we would have to go to get beyond the point where the tornado would cross the road. We knew the tornado was moving to the northeast. I thought we could stop almost in its path and collect some data with the LANL radar. We stopped momentarily, but Herb Stein, our driver, warned me that the tornado appeared to be getting wider, which likely meant it was getting closer. Herb and the graduate students urged me to get going. In a daze I agreed.

We drove on for what seemed like an eternity, shooting video of the approaching tornado from the van window, and finally arrived at about the latitude of the tornado. If it had any northward component of motion at all, we could safely collect data. When we stopped, the tornado was to our northwest, a huge cylinder silhouetted against the bright sky (Fig. 5.17). While the crew deployed the radar, I tried to set up my video camera on a tripod, but was foiled by a strong westerly wind. I braced myself against the back of the van to shield myself from the wind, which kept blowing the tripod over. I didn't realize that the Tulsa television crew did not feel safe there and had retreated even farther to the south. Coming from the north, another storm chaser, Bill Barlow, whizzed by; he had really threaded the needle. A stray hailstone hit the radar, making a pinging sound.

The radar was turned on and collecting CW data, but our crew, in their excitement, had inadvertently switched a video cable with a signal cable while setting up the radar. And as they attempted to correct the error, the tornado swept across the road to the north. So the video frames were not synchronized properly, and no signals were recorded on the approaching channel of the recorder. Fortunately, most of the wind was behind us.

We drove about a mile to the tornado's damage path where we later

5.17 *An F5 tornado, one mile away to the northwest, north of Red Rock, Oklahoma on April 26, 1991, around 6:50-6:54 P.M.*

discovered that a farmhouse had been blown off its foundation. But even before reaching the damage path, we were alarmed to see utility poles strewn in the road. They must have been hurled a half mile from the tornado—the poles alongside the road were undisturbed. So even beyond the path of a tornado, damage can occur from flying debris.

We soon learned that the tornado's path had widened to a mile. We continued on, and our path took us south a bit to Ceres and east through Red Rock, which gave its name to the tornado we had just seen, even though it didn't actually hit Red Rock but passed just to its north. Soon it became apparent that we could not catch up to it. A new storm was forming way off to the east. The TV crew, regaining courage, raced off to try to catch it, while we returned to the house that had been blown off its foundation (Fig. 5.18). We were very concerned for the lives of those who might have been in the house. No one was in sight, but we later learned that the people who lived there were safe—they had not been at home when the tornado hit. We did come across a motorcyclist, who had accidently hit a downed power line and then flipped. Herb Stein is also a paramedic and was able to assist the victim, who survived.

Another line of storms then appeared on the horizon to the northwest. Unlike the earlier, isolated storms, which had formed near the dryline, the

5.18 *The remains of a house that had just been struck by the Red Rock tornado on April 26, 1991.*

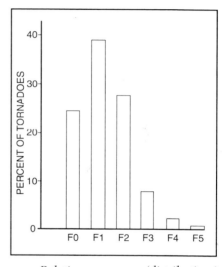

5.19 *Relative occurrence (distribution in percent) of tornadoes in the United States by F-scale. (Adapted from Schaefer et al. 1980; courtesy of the American Meteorological Society. Copyright 1980.)*

new ones lined up along a cold front and it was getting dark, too dangerous for chasing. We watched them as they spit out dazzling lightning, but decided that we had had enough excitement for one day and headed home.

Later analysis of the CW radar data showed wind speeds in the Red Rock tornado as high as 270-280 mph—in the F5 category on the Fujita scale of tornado intensity. Ours were the first measurements of wind speeds in an F5 tornado, and perhaps the highest wind speeds ever recorded on earth by a Doppler radar. F5 tornadoes are extremely rare, and F5 events account for fewer than 10 percent of all tornadoes. Most common are the weaker F0–F2 tornadoes (Fig. 5.19). Because the resolution of the small parabolic dish antennas in the LANL radar was too crude to pinpoint the highest wind speeds, we couldn't resolve an important question: whether the highest wind speeds were near the ground or hundreds of feet up.

The Red Rock tornado had been a scientific bonanza—and May, the peak month for storms, hadn't even arrived! While we missed an exceptionally photogenic tornadic storm near Laverne, caught by other chasers, it was really our only failure of the magical 1991 season, and we regained our honor several weeks later by intercepting a tornadic storm, southwest of Wichita. The annual Water Festival was in progress that day and, mindful of the April 26 tornado, people were terrified of being hit by another. The coverage we heard on the local radio was downright scary. We stopped to collect radar data of a picture-perfect tornado over a wheat field (Fig. 5.20). A crew from California with an IMAX movie camera shot dramatic footage of the tornado, which appeared, more than four years later, in *Storm-chasers*. I later saw a video on the local news in Oklahoma City, in which a ministampede of horses was initiated by the tornado—a spectacle we missed. The tornado passed mostly over open country and dissipated before arriving at Wichita. At this point we had almost completely fulfilled

5.20   *Tornado viewed to the northwest on May 16, 1991, about 5:16 P.M., near Clearwater, Kansas.*

our hopes. Yet I was not aware that we still had not collected an FM-CW data set of high-enough quality for analysis. From our CW data, however, we could determine the maximum wind speeds in four tornadoes.

What are the highest wind speeds in a tornado, and where are they found? Theoretical meteorologists Brian Fiedler and Richard Rotunno reconsidered what is called the thermodynamic speed limit. Suppose that, as first proposed by meteorologist Doug Lilly, the wind speeds in a tornado are directly linked to the hydrostatic pressure deficit associated with the vortex. The warmer and lighter the air column within a vortex of fixed width, the greater the pressure deficit, the inward-directed pressure-gradient force, and the wind speeds. The highest wind speeds associated with the hydrostatic pressure gradient represent the thermodynamic speed limit. Fiedler and Rotunno argued that this was not in fact the true limit: Air that spirals inward and upward in the vortex accelerates so rapidly that it is nowhere near a state of balance. Hydrostatic balance is not achieved and the pressure would be lower than its hydrostatic value. Indeed, when we calculated the thermodynamic speed limit of the tornadoes for which we had CW data and compared that data to the result of the calculations, we found that the limit had been exceeded.

I then decided to put my graduate students in charge of our chase van and participate in the next research flight on the P-3. On May 26 we took off from Will Rogers World Airport in Oklahoma City and headed for the Texas Panhandle, where a dryline was supposed to move eastward. Much of the day we spent flying stepped-traverse patterns—back and forth across the dryline at different altitudes. During the afternoon we noticed a storm to the north, but it appeared to be behind the dryline, in dry air. What we saw didn't agree with the aircraft data. We were missing something. We

5.21   *Dissipating tornado on September 2, 1992, several-minutes after 6:47 P.M., viewed from the south, east of Wayne, Oklahoma: (left) the tornado assumes a ropelike appearance; (right) the upper part of the condensation funnel has disappeared. Note the similarity of the figure at right to Fig. 5.4.*

continued to sample the dryline. Although we had a bird's-eye view of clouds, we had no up-to-the-minute meteorological data.

Finally the pilots reported a tornado ahead to the west (see Fig. 2.13), in the storm we had noticed earlier in the "dry air." The tornado was visible at ten thousand feet, our flight level, from forty miles away. We abandoned our dryline patterns and storms to go after this bird in the hand. Probably sensing that it was being observed, the tornado promptly died as we reached it. While we had gone off on a tangent, the storms growing along the dryline had rapidly intensified; one near Mooreland, Oklahoma, had become tornadic. Down on the ground, my crew had successfully bagged another tornado and was collecting more radar data. But by the time we arrived back at the storm, that tornado was gone.

The chase on May 26 marked a watershed: Both airborne and ground-based portable CW radar had collected data in tornadic storms. In future years, our experience in using the airborne radar would prove invaluable.

The 1992 storm season deserves mention for only one event: For the first and only time, I went on a hybrid bicycle-car chase. Late in the afternoon of September 2 I watched a storm pop up just outside my office window. I pedaled quickly home on my bicycle and negotiated with my wife to borrow her car. (Mine had just been sold and I had not yet bought a replacement.) Then I turned on the TV and saw a radar depiction of a supercell *cum* hook echo just southwest of Norman. Grabbing my camera gear, I tore out the door and headed south on the interstate. Fifteen minutes later I slowed to a snail's pace as a rotating wall cloud crossed the road overhead. Within minutes a tornado appeared, and I followed it to the east, photographing several great images (Fig. 5.21). Times like these make up for all the thousands of miles I have driven in a futile attempt to land an elusive tornado.

6

The State of
the Art

Venturam amiciat
imbrifer arcus aquam.
(The rainbow warns of an
approaching storm.)
                    —*Tibullus*

NEW INSTRUMENTS are now
being developed to help unravel the mystery of tornadoes and other severe-
storm phenomena. Some of these devices were tested in a two-year coordinated field chase program during the storm seasons of 1994 and 1995.

The Turtle, or TOTO II

TOTO was a good instrument, but it was somewhat cumbersome, and the
odds of placing it directly in the path of a tornado were low. To improve the
odds University of Oklahoma meteorologist Fred Brock put together a new
and inexpensive instrument package. Like TOTO, it was portable, but
much smaller and lighter, and data were recorded digitally, not in strip
charts as in days of yore. Housed in a metal shell shaped like a turtle, it was
in fact soon called the Turtle (Fig. 6.1). With several of these creatures
spread out along the road in an area where a tornado might cross, we stood
a chance of a hit. Yet we paid a price for the smaller instruments: They
could measure only pressure and temperature, not wind.

The new instruments made their debut in the spring of 1986. Much of
the early field work was done with University of Oklahoma meteorologist

Glen Lesins, who had chased with us earlier, and researcher Bob Walko, who was interested in tornado dynamics. Further development and experimentation by graduate students Jim LaDue and Mark Shafer led to some partial successes. On May 2, 1988, students placed a Turtle under a dissipating tornado near the small town of Reagan, Oklahoma. No useful data were obtained, however, as someone—a curious local resident, perhaps—had tampered with the instrument. Turtles were again used in 1989 and 1991. On May 26, 1991, four Turtles were deployed in the Mooreland storm. One of them, sitting, about a mile from the tornado, recorded a pressure drop of 4 mb, consistent with some of TOTO's earlier measurements. Evidence was mounting that mesocyclones are indeed associated with pressure drops of around 5 mb.

Turtles were modified in 1994–95 by University of Oklahoma meteorologist Jerry Straka and graduate students Mike Magsig and Frank Gallagher, and used in Project VORTEX (Verification of the Origins of Rotation in Tornadoes Experiment), conducted in the southern plains. There were, however, no direct hits by tornadoes.

On June 8, 1995, a pressure drop of more than 50 mb was recorded by a new instrument, similar to the Turtle, left in the damage path of a large tornado near Allison, Texas (Fig. 6.2). The full meaning of this measurement made by physicist Bill Winn and his group from New Mexico Tech is not

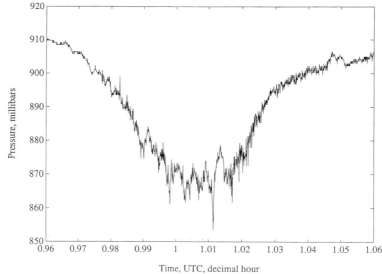

6.1 *(left) A Turtle being displayed by University of Oklahoma undergraduate student Mike Magsig in front of the Energy Center at the University of Oklahoma in Norman, September 1993.*

6.2 *(right) Pressure measurements as a function of time made by Bill Winn, Steve Hunyady, and Grayden Aulich from the Langmuir Laboratory for Atmospheric Research at New Mexico Tech in Socorro, on June 8, 1995, in a tornado near Allison, Texas, in the eastern Texas Panhandle. The pressure minimum took place over a five-minute period; there is one brief dip in pressure of about 15 mb in a few seconds. A windmill was found blown down to the south near the instrument.*

yet known. If it is correct, then pressure drops in tornadoes may indeed be comparable to those in hurricanes and typhoons, but over much shorter distances.

A Pack of TOTOs: The Mobile Mesonet

Just as Turtles or similar instrument packages can be dropped off in or near the path of a tornado, the chase vehicle can be positioned to take *in situ* measurements. In the spring seasons of 1994 and 1995 during VOR-TEX, instruments similar to those in TOTO, but updated to record pressure, temperature, and wind digitally, were mounted in an armada of sixteen cars that we called the "mobile mesonet." The armada had the advantage of mobility: the instruments did not have to be retrieved like the Turtles but remained with the chase vehicles. On the other hand, the chase vehicles did not want to remain in the path of an oncoming tornado! An innovation of Jerry Straka, and of Erik Rasmussen and engineer Sherman Frederickson, was use of a network of Global Positioning System (GPS) satellites, which sent telemetry regarding the positions of the individual cars to the field coordinator's van, where strategies for observing were mapped out.

During the two-year experiment the armada experienced some wild weather. On May 6, 1994, 80-mph winds rocked the field coordinator's van. And in late May the following year, hailstones pounded and severely damaged some of the cars, smashing windshields and side-view mirrors. Rear windows were completely obliterated by hail and large hail left car bodies with dents the size of grapefruit (Fig. 6.3). In June, high-quality observations were made near several large tornadoes in west Texas and the Texas Panhandle. Although the mobile mesonet requires a substantial network of roads to accommodate a fleet of sixteen vehicles, the VORTEX project's success at collecting data near tornadoes suggests that the concept holds promise for future field experiments.

In addition to the mobile mesonet, the Oklahoma mesonet, a system of automatic recording weather stations on small towers in every county, was designed and installed in the early 1990s, mainly through the efforts of meteorologist Ken Crawford of the University of Oklahoma. By the time of VORTEX, all these stations were up and running, transmitting data in real time every fifteen minutes. (Other data were archived every five minutes, but available to researchers only after the fact.) So far, although some straight-line winds as high as 100 mph have been measured, there have been no direct hits by a tornado.

Remotely Piloted Vehicles

Another way to make measurements near tornadoes is to fly an instrumented drone, controlled from the parent chase vehicle. If remotely controlled probes have worked in such harsh environments as volcanoes, the moon, and other planets, then they should work right here on earth, in tor-

6.3   *The VORTEX armada of mobile mesonet vehicles lined up and ready for duty at NSSL. Bob Davies-Jones is proudly pointing out his war wounds, hail damage suffered the day earlier in a hailstorm in north-central Kansas on May 12, 1995. (Photograph by Frank W. Gallagher III.)*

nadoes. As early as 1963 meteorologist Fred Bates suggested that drones could be used to probe tornadic storms. However, the suggestion must not have been taken very seriously, for it wasn't until 1988 that aerospace engineer Karl Bergey and his students at the University of Oklahoma designed a remotely piloted vehicle (RPV) for use in severe storms.* They built two planes, with five- and eight-foot wing spans. One was equipped with a video camera for viewing real-time images. Bergey has since retired, the project ended, and no thermodynamic measurements were ever attempted. However, meteorologists Nilton Renno and Earl Williams used their own instrumented RPV in Florida and Arizona to make measurements in clear-air thermals and in small cumulus clouds. Renno, Williams, and I would like to make measurements near tornadic and potentially tornadic storms someday with their RPV, but so far adequate funding has not yet materialized.

Helicopters as Observation Platforms

Can helicopters be used to study tornadoes? To date, most observing has been done on the ground. But it is clear that helicopters, since they can roam anywhere and hover, are better than aircraft, which cannot pass by a tornado under a certain speed, or ground-based vehicles. On May 22, 1981, a TV crew in a helicopter filmed a large tornado west of Oklahoma City. It was a daring and dangerous operation conducted on the spur of the moment, but it produced a breathtaking video. On July 18, 1986, another TV crew in Minneapolis, not out to chase tornadoes, serendipitously obtained a breathtaking video of a multiple-vortex tornado uprooting trees. And in Colorado, a TV helicopter shot action-packed videos of tornadoes in Denver on June 15, 1988, and near Fort Morgan on May 30, 1996.

During August 1993, with the support of NOAA and National Geographic Television, Joe Golden and I flew for a week in an NOAA helicopter in the Florida Keys to film movies of waterspouts. The movies were to be used in a National Geographic television special on cyclones, but we also hoped to use the data to estimate wind speeds. We did see about a dozen waterspouts† (Fig. 2.23), including some that produced spectacular vortices of sea-surface spray. Our waterspout expedition impressed me with how common waterspouts really are; the frequency of their occurrence is underestimated. I was also impressed with how useful the helicopter is for hovering near vortices such as waterspouts. The main problem using them to study tornadoes, is that large hail or airborne debris could pose a danger. In addition, helicopters are expensive to operate, and have to refuel frequently.

---

* Incidentally Bruce Sterling, in his 1994 science fiction novel *Heavy Weather,* discusses a futuristic drone for making measurements in tornadoes.

† Although the footage unfortunately could not fit in the TV program because there was not enough time for it, it was included in a video called *Cyclones,* later marketed by the National Geographic Society.

Electrical Effects and Tornadoes

The relationship, if any, between thunderstorm electricity and tornadoes is still not clear. While studies by NSSL scientist Bill Taylor in 1973 and 1974 showed that about 80 percent of all tornadic storms produce an unusually large number of lightning flashes, most of them were probably cloud-to-cloud flashes, not cloud-to-ground flashes. In fact, I have observed many tornadoes near which very few cloud-to-ground lightning flashes, if any, were visible.

The interest in electrical effects in tornadoes probably stems from the unusually large number of high-frequency sferics (electromagnetic noise radiated by lightning) detected in a tornadic thunderstorm in Oklahoma back in 1950 by electrical engineer Herbert L. Jones, who was monitoring radio waves.

Other, more recent studies have produced mixed results. Perhaps the most widely publicized work is that of Newton Weller, an electronics expert who in the late 1960s devised a method that, he claimed, could be used by TV viewers to detect nearby tornadoes: Turn the dial to channel 13 and adjust the brightness control until the screen is nearly black; then turn the dial to channel 2, not resetting the brightness control. Lightning flashes will then appear on the screen as actual flashes. But if the screen becomes bright and remains bright, a tornado is likely within twenty miles. I don't understand why this should be so and don't believe it works.

Weller claimed this system works because channel 2 operates at 55 MHz and is closest to the frequency of what he calls a "tornado pulse." It is not clear if this is actually the case. Weller also advised, "Don't get so engrossed in watching the TV screen that you forget to seek cover fast when a tornado is approaching." Obviously, the Weller technique does not always work, and one should probably depend on the more reliable tornado warnings issued by TV stations rather than signals visualized on television. In any event, the Weller technique is certainly not applicable to cable TV.

After a powerful tornado struck Plainville, Illinois, in August 1990, meteorologist Anton Seimon discovered, using data from the National Lightning Detection Network, that the charge of cloud-to-ground flashes switched abruptly from positive to negative as the tornado began. Lightning data from other storms, analyzed by Don MacGorman and coworkers at NSSL, do not show such a clear-cut relationship. The NSSL team found that some particularly violent tornadoes occur in storms that produce cloud-to-ground flashes that lower negative charge, but few that lower positive charge. They also found that the charge lowered by ground flashes can switch in storms that produce only weak, if any, tornadoes. There seems to be no clear way to use flash polarity to detect tornadoes.

Atmospheric physicist Bernard Vonnegut in 1960 proposed how tornadoes could be driven electrically. Elaborating on the idea proposed in *De Rerum Natura* by the Roman philosopher Lucretius in 60 B.C.E. that "electrical energy could heat a volume of air to such a high temperature that abnormally intense convection would take place," Vonnegut suggested three processes that could account for the heating: lightning, "glow discharge," and the movement of ions in an electrical field. Another mechanism had been proposed by physicist R. Hare in 1837: "highly charged air could be accelerated to a high velocity in a strong electric field." According to a study by Don MacGorman, the latter couldn't work because charged water drops and droplets are too massive to move quickly, and charged air molecules (free ions) don't drag the rest of the air along with them.

The current thinking is that the observed and postulated supersonic wind speeds in tornadoes can be accounted for without resort to electrical theory. However, we must remain open to the possibility that in the most powerful tornadoes electrical effects could play a role, perhaps in supplying a wind perturbation that could be amplified dynamically.

Mobile Doppler Radars in VORTEX

The NOAA P-3 airborne Doppler radar, first used to study severe thunderstorms during COPS in 1991, made a return in 1994. The radar still hadn't collected wind data in a visually confirmed tornado—back in 1991 it had arrived at storms just as the party was ending. But on May 25, 1994, it came close. We were on the ground with our portable FM-CW/CW Doppler radar, and aloft the P-3 flew flight legs by a tornadic storm in a sparsely populated region near Northfield, Texas. Then, ten to twenty minutes before a tornado appeared to us on the ground, other storms moving in from the southwest forced the P-3 to break its flight pattern. The weather gods prevented it from flying by the storm again, and collecting tornado data, but the LANL radar, finally, acquired a good FM-CW data set. A mapping at 250-foot increments of the Doppler wind spectrum revealed an unprecedented look at the structure of a tornado (Fig. 6.4). Wes Unruh's software and painstaking data sampling and analysis finally paid off. Analysis of the larger-scale (too broad to resolve a tornado) windfield of the

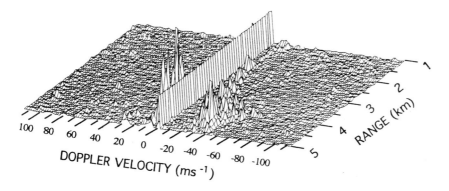

6.4 *FM-CW Doppler wind spectra as a function of range in a tornado on May 25, 1994, near Northfield, Texas. The radar's view was centered to the left of the tornado. The "fence" at zero velocity has been inserted for display purposes. The vertical scale, which is linear, is the spectral density. Approaching (receding) velocities are negative (positive). Note that the maximum winds that appear above the noise floor are approaching at least as fast as 60 m/sec at a range of just over 4 km, which is in the tornado's damage path. (Courtesy of Wes Unruh.)*

6.5 *The NOAA P-3 aircraft outfitted with an airborne Doppler radar, at Will Rogers World Airport in Oklahoma City, photographed on June 8, 1995, prior to operations. The antenna system is mounted in the tail of the aircraft. The National Center for Atmospheric Research's Electra aircraft outfitted with ELDORA, Electra Doppler Radar, looks nearly identical.*

Northfield storm, based on data from the P-3 radar, showed circulation along a large hook echo that probably evolved into the tornado.

Several days later, on May 29, I flew in the P-3 (Fig. 6.5) near an outflow boundary southwest of Wichita Falls. We flew patterns by a line of cumulus and towering cumulus, and time slowed down as we waited for the clouds to mature, we hoped, into tornadic thunderstorms. The watched tornado pot was slow to boil, but eventually it did, and the challenge now was to determine which storm was most likely to succeed.

To the northeast near Archer City, where the movie *The Last Picture Show* was filmed, there was an incredibly vigorous-looking storm (Fig. 2.20)—and within an hour it did produce a tornado. But the ground teams from VORTEX were working the next storm down the line, near Newcastle, Texas. We began our flight patterns at or near the cloud base. During COPS we had flown much higher, at about ten thousand feet. By staying near the cloud base, we would be underneath potentially turbulent areas and could get closer to the tornado, possibly even getting a glimpse of it. Yet we would lose the ability to analyze the horizontal wind field at high levels. From close range, high levels in the storm are seen by looking up at angles greater than forty-five degrees. At such steep angles, the Doppler radar sees mostly vertical velocity, not horizontal velocity.

Finally, on a flight leg that took us toward the west, I spotted a large tornado off to the north. Owing to haze, precipitation, and the angle of the sun, the tornado was not sharply defined and was difficult to photograph. But it was a big one, and I was surprised that the ground crew, whose conversation we were monitoring, did not sound excited. How could they miss it? The reason: They were observing cloud features associated with a hook echo to the east, and the tornado formed to the *west* of the echo and mesocyclone. Roger Wakimoto's analysis of the single Doppler radar from the flight showed that the tornado formed from below, as landspouts apparently do. Fuel was thus added to the controversy: Do tornadoes begin as mesocyclones that build down or as independent circulations that build upward?

On June 8 we had another tornadic flight in northeastern Colorado while the VORTEX ground crews focused on a surface low-pressure system in Oklahoma. Then we heard about tornadic storms in western Nebraska, and headed up there immediately. Having spent so much time on the ground in ridiculously long and boring drives, being able to jump states in a plane was a delicious treat.

Storms are always windier in the other fellow's yard, and once in Nebraska, we found that the storms billed as tornadic had evolved into a large squall line with no apparent potential for supercells. We flew flight legs, collecting data, for perhaps an hour or so. Then I spied an isolated cell on the surveillance radar display and pleaded to divert the plane to that cell. Just as we reached it, we could see a tornado about to pass into its dissipating stage. We flew past it and I took photographs in rapid succession; these can be viewed side by side in three dimensions as stereo images. In the seconds between photographs, the tornado changed little in appearance, but the plane moved a sufficient distance so that the viewing angle changed noticeably. We flew past the storm and collected data for an hour and a half without another tornado forming.

Millimeter-wavelength Radars: Ultra-high-resolution Probing

To achieve high spatial resolution at 3-cm wavelengths, a large antenna must be used. Neither the LANL nor the NOAA radar system can accommodate a much larger antenna. In order to effectively map the windfield in a tornado, which is three hundred to three thousand feet across, resolution on the order of thirty to three hundred feet is required. The LANL radar's resolution is eight hundred to a thousand feet at an operating range of approximately two miles, and that of the airborne system is a thousand to thirteen hundred feet at an operating range of six miles. Getting any closer to the tornado is both difficult and dangerous.

The solution is to use a radar that operates at a higher frequency with a smaller-beamwidth antenna. Millimeter-wavelength radars, for example, have been around for some time and were originally used for studies on cloud formation and behavior. They are more sensitive than the centimeter wavelength radars to backscatter from tiny particles like cloud droplets and ice crystals.

In 1992 I began to work with electrical engineer Bob McIntosh and his group from the University of Massachusetts to map the windfield in tornadoes, using one of their 3-mm wavelength systems. This system can push the envelope of resolution. An antenna only twelve inches wide has a beamwidth of 0.7 degrees and yields a cross-beam resolution of about a hundred feet at a range of two miles. An antenna three feet in diameter can yield an amazing thirty-foot resolution at the same range. Currently, it can "see" as far as one mile in clear air, and further development is expected to increase its range to three miles or more.

The range of the millimeter-wavelength radar, however, is limited because of attenuation from intervening precipitation. Another problem is that the Doppler dilemma is accentuated. Using traditional pulsed methods, the maximum unambiguous velocity is only 27 mph with a maximum unambiguous range of six miles. McIntosh's group implemented an ingenious scheme first proposed by radar meteorologist Dick Doviak and engineer Dale Sirmans at NSSL back in the 1970s. They used interleaved sets of polarized pulses to detect wind speeds unambiguously as high as 178 mph at ranges beyond six miles. Polarized radar pulses can be thought of as analogous to horizontally and vertically polarized light beams. When, for example, horizontally polarized light hits a lens that admits only vertically polarized light, it appears dark. Thus, the radar can "listen" separately to backscattered signals from vertically and horizontally polarized pulses.

McIntosh's students designed and built the system that was installed in our chase van. We had the University of Oklahoma motor pool cut a hole through the roof of the van so that the antenna could poke through like a periscope (Fig. 6.6). The scans are programmed by computer. When we set up, load levelers are lowered to the ground like the legs on spacecraft landers on the moon and Mars. It takes only a few minutes to set up and start collecting data.

Supported by NSF, we were able to initiate the system in the spring of 1993, but we intercepted only two tornadoes. The first one, on May 7, appeared briefly a short distance from Wellington, Texas. By the time we got the antenna up, the tornado had disappeared behind heavy rain. On the next day we intercepted a tornado south of Wichita Falls, but it too disap-

6.6 *The University of Massachusetts mobile 3-mm wavelength Doppler radar system mounted in a van from the University of Oklahoma, on June 7, 1993, near Elmo, in central Kansas.*

peared behind precipitation before we could start to collect data. Beginner's luck was not to be ours—our data collection with the new radar was limited to nontornado events.

Our efforts in 1994, the first year of VORTEX, were just as frustrating. In May a freelance writer from *The New York Times Magazine* and a freelance photographer accompanied us three hundred miles all the way to eastern New Mexico in the hope of seeing some severe weather, which never materialized. On the same trip the crew shooting the IMAX movie *Stormchasers* also followed us in a futile attempt to film more tornadoes. Our only chance to probe one came on May 25 near Northfield, Texas. Although we were successful in collecting data with the LANL radar, electrical problems plagued us and we were not able to collect data with the millimeter-wavelength radar.

In 1995, during the second year of VORTEX, our frustration mounted. On April 17 a small tornado crossed the road only a few hundred feet from us in south-central Oklahoma and we were without the radar—which was also being used in other experiments and on that day was on its way down from Massachusetts. But we consoled ourselves with the thought that even if we had had it, there wouldn't have been enough time to set it up safely at our close range. Our compensation was a beautiful tornado. Instead of hearing a roar like a freight train, we heard a soft, swishing sound as rain whipped against the side of the van from the west. We were so close to the tornado that the whitish-gray condensation funnel took on an eerie, ghostlike appearance. The edge of the funnel was blurred, not sharp, as it usually appears from a distance. It was an amazing experience that ranks up there with the Red Rock close encounter.

Our participation in VORTEX gave us the opportunity to collect data from many platforms, including the Mobile Mesonet, the NOAA airborne radar, Turtles, and M-CLASS systems from NSSL and NCAR. The real-time nowcasting information we received was superb; we no longer had to rely on NOAA weather radio warnings, spotter reports, and local radio broadcasts. Field coordinator Erik Rasmussen used his cellular phone almost continuously, and he and his team radioed up-to-the minute information to all the chase vehicles. There was frequent militarylike chatter, and at times our conversations sounded like space-launch talk. For example, it was common to hear conversations such as "Probe 1, this is the FC: Deploy your turtles. FC, Probe 2: Location 36.521 N . . . Proceeding to the intersection of . . ." We got used to hearing the sequence of beeps containing information about the GPS-determined locations of the caravan. We can easily get our own detailed Doppler-radar information from the comfort of our chase van on a cellular phone connection between a laptop computer in the van and the computer back home. But this can disrupt our timing. We could go with the flow, sticking with the VORTEX caravan. Or we could go where we please, possibly missing out on the chance to get supporting data. We have tried both ways.

On April 26 we were good soldiers, dutifully following the caravan— and missing a tornado. The armada was en route to Texas to intercept tornadic storms south of the Red River. North of the river, a storm with a mesocyclone appeared, and we deviated from our route to intercept it. I had doubts about the storm—it was north of a warm front and embedded in a cool air mass at the surface. It reminded me of the "storm before the storm" of April 10, 1979, and so it was—nothing came of it. Once back on our course, several members of the caravan had to stop to refuel. We waited impatiently. Finally I couldn't stand it anymore and got permission to barrel ahead, even though we knew that when we were twenty miles from the field coordinator, we would lose radio contact. When we arrived in Gainesville, Texas, debris was scattered across the interstate highway and on either side of it—a tornado had crossed the road minutes before.

On May 15, 1995, during the second year of VORTEX, no VORTEX operations were scheduled. Yet we believed there was the possibility of activity in the Texas Panhandle, and took off by ourselves. When we arrived at Amarillo, storms were building up to our northwest and south. We chose an isolated one to the northwest because it was visible and in our target area, while to the south several storms were lined up, and we would have to core-punch them to get to the southern end. A tornado was reported in the southern storm, but none elsewhere. Dejected, we spent the night in Plainview. The next morning, accessing the University of Oklahoma computer, we learned that there was a chance of activity both right where we were and way up in southwest Kansas, a four- to five-hour drive away. We weren't all that sure that the Kansas scenario would play out. In fact, we weren't sure the Texas scenario would work out, either, because a thick, high shield of clouds had overspread our area. We don't know exactly why high clouds are a negative indicator, but we suspect they modify solar heating, so the formation of storms is less likely. We chose to stay put, even though the VORTEX armada, having begun in Norman, was on its way to southwest Kansas. The cloud shield persisted until late afternoon, while tornadic storms formed in southwest Kansas, all for the benefit of the caravan. We had gambled and lost.

On the next day, a large outbreak of tornadic storms occurred but not near the VORTEX armada, which we had now joined. As we targeted north-central and eastern Oklahoma, tornadoes were striking northwest Oklahoma. But the 3-mm radar finally collected useful data in two counterrotating vortices in a supercell near Locust Grove. Although the vortices were not tornadic, they were only fifteen hundred feet across and were accompanied by mirror-image hook echoes. Such features were not detectable by a nearby WSR-88D radar because they were too small.

On May 22, in the Texas Panhandle near Pampa, we watched a tornado in the distance run through its life cycle. No roads led to the tornado, which was at least ten miles away, so we watched helplessly, unable to collect data. Other storms that day failed to produce tornadoes (Fig. 6.7). About a week later we shipped the radar back to Massachusetts. Andy Pazmany, the radar guru, without whom we could not operate the radar with much confidence, had to return there by June 1, and we felt that, based on both the long-term average weather pattern and the existing weather pattern, it was unlikely that more severe storms in Oklahoma or surrounding areas would crop up. As it turned out, we were bearish in what proved to be a bullish tornado market.

6.7 *The monster hailstorm of May 22, 1995, viewed from the east, south of Shamrock, Texas, in the Texas Panhandle. Blowing dust near the ground is being sucked up into the updraft from the left at 7:05 P.M.*

ELDORA's Debut in the Southern Plains

During the second year of VORTEX another airborne radar was used near severe thunderstorms. Known as ELDORA (Electra Doppler Radar), the new 3-cm radar was mounted in the NCAR Electra aircraft. It operates like the NOAA radar, but is more sensitive (more pulses transmitted at slightly different frequencies are averaged) and, since its antenna spins more rapidly, has better resolution along the flight track. The Electra is similar to the P-3, except that it is not as hardened for turbulent flight. The P-3 has flown safely through dozens of hurricanes; the Electra has not yet been put to the test. Using a dual-PRF processor, ELDORA can achieve a maximum unambiguous velocity of 178 mph, while the P-3 radar's maximum unambiguous velocity is only 29 mph. The low folding velocity in the P-3 data creates quite a headache for researchers, who must unfold high-velocity data in severe thunderstorms.

ELDORA had made its debut during the winter of 1993 in the South Pacific during a large international field experiment called TOGA-COARE (Tropical Ocean and Global Atmosphere—Coupled Ocean-Atmosphere Response Experiment). Certain hardware problems cropped up, and Roger Wakimoto and I, co-principal investigators of the 1995 experiment, hoped that the bugs had been removed in time for VORTEX. ELDORA's first major success came on May 16, while my crew and I were on the ground in West Texas with the millimeter-wavelength radar, waiting for storms that never came; ELDORA, with Roger at the helm, was flying in southwest Kansas, collecting data near tornadic storms.

No sooner had the Massachusetts radar returned to its home than good conditions for tornadic thunderstorms, not as often present in June as in May, appeared in the Texas Panhandle. I decided to fly in the Electra with ELDORA for most of the duration of VORTEX and put my graduate students in charge of the LANL radar.

On June 2 we took off for the Texas–New Mexico border. Thunderstorms popped up just as we arrived. Strong surface winds independent of

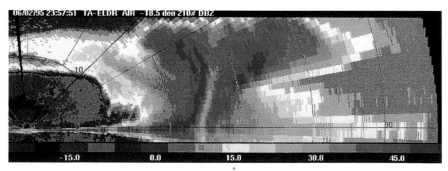

6.8 *Weak-echo hole as seen by ELDORA on June 2, 1995, in a tornadic supercell in West Texas: a nearly vertical slice through the storm. Reflectivity is shown in dBZ. The weak-echo hole extends from just above the ground up to the top of the storm, which is nearly 10 km up. (Courtesy of W.-C. Lee, Atmospheric Technology Division, National Center for Atmospheric Research.)*

the storms had stirred up a blinding dust storm. Soon several tornadic storms formed, each positioned so that it was not possible to fly legs near them. Owing to other storms just to their south, we had to maintain a distance of twelve miles from the west side of the storms. Flying at only a thousand feet above the ground, we were tossed about mercilessly by turbulence in the cold surface outflow air mass. We probed the storms blindly, unable to see the tornado or to get closer. On the ground, the VORTEX crews were watching two impressive, large tornadoes near Friona and Dimmitt, Texas, southwest of Amarillo; aloft I observed them as spectacular radar images, showing holes extending from low levels up to the top of the storms (Fig. 6.8). The holes, about a mile in diameter, probably indicated either air moving rapidly *upward* or precipitation being centrifuged outward. (Hurricanes also have echo holes in their eyes five to fifty miles in diameter; air moves *downward* in hurricane eyes.) We could see some of the damage inflicted by the Friona storm—a freight train knocked over, some building and tree damage.

The flight of flights wasn't until a week later, on June 8. We flew out of Oklahoma City early in the afternoon and headed, at an altitude of about ten thousand feet, for the Oklahoma Panhandle. There a supercell had formed at the intersection of a front and a dryline. The ground crews deployed the mobile mesonet in strategic places around the storm. On our way I had noticed other storms firing up along the dryline, which extended south through the Texas Panhandle, and suspected that we would have to check them out. Severe thunderstorms frequently form at the intersection of the dryline and a front, but on a day like this, with the low-level air mass very moist and hot, storms along the dryline could also be widespread and severe. But first we were obligated to probe the storm the ground crews were observing.

We dropped down to cloud-base level to run our flight legs by the storm. The features there were not yet well defined, but eventually they took on the appearance of cold outflow, and I became pessimistic about our chances of catching a tornado. Something really interesting, I felt, was happening off to the south, where we could see sharply defined anvils. We stayed with the storm for two hours, what seemed like an eternity, without any knowledge of those storms.

Finally the field coordinator gave in. We flew south under a veil of mamma hanging from the anvils, passing several storms on the way. The one most likely to produce tornadoes, I believed, would be the southernmost. As it would turn out, it really didn't matter which storm we picked, for most of them eventually spawned tornadoes or were adjacent to ones with tornadoes.

Visibility was severely limited in the hazy, moist air. We dropped to a thousand feet, marveling at the crisp back edge of the anvil overhead. As we executed our flight pattern, we heard about tornadoes in the storm we were looking at, the southernmost one, but could see nothing clearly—from the west, sunlight was scattering off the low-level haze at a low angle, making it nearly impossible to see a tornado if there were one. As depicted on the airborne radar the storm was breathtaking, showing a beautiful,

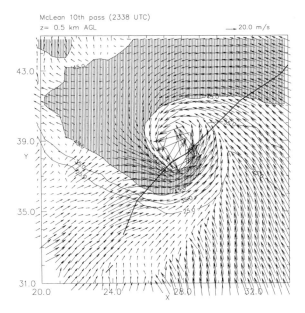

6.9 *Pseudo dual-Doppler analysis of the ground-relative wind field of a tornadic supercell on June 8, 1995 in the Texas Panhandle. Data are from the NCAR ELDORA. Radar reflectivity factor (dBZ, 45 dBZ contour is shaded) is contoured by solid lines; wind field depicted is at 500 m above the ground, at 6:38 P.M. The damage paths of two in a series of tornadoes are depicted by thick solid lines; the longer one, coincident with the hook echo, was associated with a tornado so violent it ripped pavement out of the ground. Highest wind speeds resolved are around 80 m/sec (almost 180 mph); the actual highest wind speeds in the tornado were probably much higher. The x and y distances are given in kilometers from Clarendon, Texas. (Courtesy David Dowell.)*

well-developed hook echo (Fig. 6.9). We were flying past a tornadic storm and couldn't see the tornado!

After flying patterns by this storm for over an hour without seeing a tornado, we finally came close enough to the cloud base so that I could make out, just barely, the outline of what appeared to be a huge tornado. We crept up even closer to the storm, incredibly close, only about three and a half miles from it. There appeared the silhouette of a huge tornado. We could see the ground being churned up below us, with dirt and vegetation flying about. It was a truly magnificent, awe-inspiring sight. On each flight leg, the view of the storm switched from starboard to port and back again. In between flight legs we were jumping back and forth between windows on either side of the aircraft. Roger Wakimoto suggested that we ask the pilots to ease away a bit, since we were awfully close to the tornado. I felt that since we hadn't run into any turbulence, we could safely hold our course.

Several legs later, we hit an unexpected bump at a thousand feet, and dropped nearly instantly to five hundred feet above the ground. Those not under the clutches of a seatbelt, including myself, hit the ceiling. Luckily, I had stowed my camera equipment in its padded case. I had learned early on, as a graduate student observer in a cloud-seeding flight over central Florida, that you had to be very careful not to hold on to loose equipment during turbulence. Several others on the flight were not so lucky; they received cuts and bruises and were extensively bloodied. We turned back to Oklahoma City for medical attention. Luckily, no one was seriously hurt. But excellent airborne Doppler data had been collected for much of the lives of five tornadoes—a real gold mine for researchers.

The ground crews had arrived in time to collect mobile mesonet and turtle data near the great tornado we had seen, the Kellerville tornado, and ones that followed. The last tornado of the day, near Allison, Texas, grew to more than two miles in width (see Fig. 6.2 for pressure measurements in this tornado). Some of the ground crew were stranded when they couldn't refuel after their vehicles ran out of gas. The tornadoes had cut off power to sections of the Texas Panhandle, including gas stations. It was a dramatic end to VORTEX.

DOW: *The Doppler on Wheels*

Josh Wurman, one of our faculty members who had worked with Doppler radars at NCAR, approached me with a new idea: Why not mount a 3-cm pulsed Doppler radar in a van or truck? I had considered doing this, too, but had some doubts about the mobility of such a system. The antenna would have to be much larger than either of those we had been using, and I did not have the time or experience to devote to building such a system; others had handled all the design and engineering aspects of the LANL and 3-mm radars.

From NCAR Josh managed to procure a spare high-powered, pulsed 3-cm radar transmitter and a Doppler signal processor and, with his coworkers, managed to fit the system into one of NSSL's older trucks. A military 1.2-degree-beam antenna, once used to track missiles, was mounted on the bed of the truck (Fig. 6.10). Josh's team worked hard to get the DOW

6.10 *The first Doppler on Wheels (DOW) and Josh Wurman, in Norman, Oklahoma, on August 30, 1996.*

(Doppler on Wheels) up and running before VORTEX ended. It was only five months or so from the time parts had been ordered until the first use of the DOW on May 12, 1995, in a big nontornadic hailstorm in Kansas.

On May 16 the DOW hit pay dirt. A tornado was probed from a range of five miles near Hanston, Kansas. The data showed an echo hole inside the tip of a hook echo, suggesting that radar-reflecting rain, hail, or debris were being centrifuged outward. Although the velocity data were severely folded, it was possible to unfold the velocities, obtaining the first mapping of a tornado's wind field. Yet the highest velocities were still unresolved, owing to the limited horizontal resolution at the five-mile range. It was a marvelous application of known technology, because the entire system was still compact enough to be moved around.

The best, however, was yet to come. On June 2 the DOW recorded data at a range of only two miles in a large tornado near Dimmitt, Texas (Fig. 3.18), and on June 8 it collected even more data in the Texas Panhandle. A new era in mobile Doppler radars was here. Now, with the DOW able to sample entire storms and nearby tornadoes with moderate resolution, the 3-mm radar able to achieve high resolution nearby, and the ELDORA and P-3 radars able to follow storms around, we could map the wind field below the cloud base—both while a tornado was forming and maturing, and while it was dissipating.

The next step was to coordinate two or more mobile ground-based radars around a storm for further coverage. An upgraded, more powerful version of the DOW, the DOW2, was constructed. It is capable of achieving a higher maximum unambiguous velocity, and has an antenna with a 0.9-degree-beamwidth antenna. In March 1996, the older DOW was upgraded also, and now triple-Doppler coverage is possible, using both systems and the 3-mm radar. In 1996 the DOW collected data in a tornado in Kansas and in 1997 more data in several tornadoes in Oklahoma during a small field program dubbed SUB-VORTEX. A major milestone was reached on May 26, 1997, when dual-Doppler data were collected in a tornado in northeast Oklahoma. It is ironic that the use of two 3-cm DOWs marked a return to the original 3-cm systems used back in the late 1960s. Since that time the problem of achieving higher maximum unambiguous velocities has been solved. Furthermore, attenuation is not a problem when the feature being probed is nearby. The future use of 10-cm Doppler radar systems like those used by NSSL in the 1970s and 1980s may therefore be limited to the fixed-site, operational radars used by the National Weather Service and other government agencies.

Doppler Lidars

Another branch of the radar family is the *lidar,* a laser radar that emits light at much higher frequencies and much shorter wavelengths than a microwave radar. A lidar has the virtue of being able to sense backscatter from atmospheric aerosols in clear air. An ordinary radar senses clear-air return when there are insects or sharp gradients in air density. Index-of-refraction (a measure of the degree to which a beam of radiation is bent)

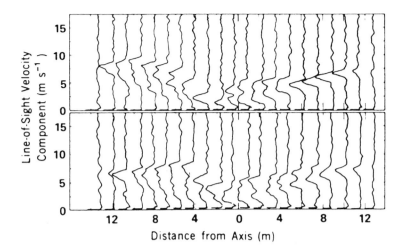

6.11 *Doppler spectra collected by a Doppler lidar on August 26, 1976, north of Key West, Florida, during two aircraft passes (top and bottom) by a waterspout, made approximately three minutes apart. The aircraft was moving from right to left. The Doppler lidar was pointed perpendicular to the flight track. The departures from a straight vertical line are relative backscattered power at the velocity indicated on the ordinate. From the figure we find that return from the lidar began just under twelve meters from the center of the waterspout and that the maximum wind velocity just under 10 m/sec was located near the edge of the condensation funnel. (From Schwiesow et al. 1981; courtesy of the American Meteorological Society. Copyright 1981.)*

gradients occur where air density, determined by the temperature and humidity, varies. The speed of the radar's radiation varies with the index of refraction. When radiation encounters a rapid change in index of refraction, the path of the beam changes and a small portion is backscattered. Sharp index-of-refraction gradients occur along air-mass boundaries such as fronts, the dryline, and outflow boundaries, and at the edge of thermals. A lidar is not restricted to these conditions. In addition the diameter of a light beam emitted by a lidar is only an inch or so wide, far narrower than the radar beam, and does not spread as much with range.

Yet a lidar has drawbacks, too. It cannot penetrate very far into clouds or precipitation; owing to the length of the pulses, its resolution in the along-the-beam direction is not any better than a radar's;* and it is more difficult to operate if the platform on which it is mounted is being bounced around. A powerful lidar is also expensive to build and operate.

Back in August and September of 1976 in the Florida Keys, NCAR scientist Ron Schwiesow and his collaborators, using an airborne CW infrared Doppler lidar from NOAA, were able to obtain Doppler wind spectra with 2.5-foot resolution in the cross-beam direction in waterspouts (Fig. 6.11).

---

* This is true for conventional lidars. The military may have more powerful ones with shorter pulse lengths.

6.12 *Steam devil over Lake Thunderbird, east of Norman, Oklahoma, on December 22, 1989, at about 11:00 A.M. Arctic steam fog or "sea smoke" is seen hovering over the lake to the east. The air temperature was about 0°F, while the lake temperature was well above freezing.*

During the late spring of 1981, Dick Doviak, Bill McCaul, NASA scientist Dan Fitzjarrald, and I had at our disposal a lidar aboard a NASA Convair 990 aircraft. The scanning strategy was identical to the one implemented ten years later on the NOAA P-3. On June 30 we flew in central Oklahoma near gust fronts and around towering cumulus clouds, some of which were developing into thunderstorms. Unfortunately, the air-conditioning system on the aircraft failed, and when we flew near cloud-base level, where the moisture content was extremely high, condensation formed on the laser, which led to a short circuit and an abrupt shutdown of the system. Back on the ground, some NSSL scientists who happened to be at a bank in downtown Norman, were watching a small, weak tornado spawned by intersecting gust fronts. What poetic justice!

The next year Doppler ground-based lidars from NOAA and NASA employed during the Joint Airport Weather Studies (JAWS) project provided data that were used to produce analyses of gust fronts. Our hopes were high. We would use the lidar system again in the southern plains, during the heart of the storm season. But in August 1985 fire destroyed the expensive NASA aircraft and the lidar while it was on the ground, and we haven't had the opportunity to use one since. NCAR later began to develop NAILS (NCAR Airborne Infrared Lidar System), which we had hoped to use, especially in the study of such smaller atmospheric vortices as waterspouts, steam devils, and dust devils. But as of today, the NAILS is not yet available.

Steam devils (Fig. 6.12) are vortices about three feet in diameter, twenty feet or more high, and look like tiny waterspouts. They can form over relatively warm bodies of water on days when the air is at least 68°F cooler than the water, creating very unstable conditions. Most do not have a cloud base above them. Steam devils are common over the Great Lakes during the early winter and off the coasts of North and South Carolina when cold continental air blows out over the Gulf Stream. They can even be seen on small lakes: On days when the temperature is below 0°F and the sky is clear, I have gone out to Lake Thunderbird, near Norman, to chase them, with considerable success. I have even seen steam devils rising from a frosty golf course when the sun is heating the ground (Fig. 6.13).

Dust devils are narrow vortices, only about three to ten feet across (Fig. 6.14), that tend to appear on days when the lapse rate near the ground is steep. Common over desert areas of the southwest United States and the Great Basin, and other deserts of the world, usually when no clouds are overhead, they are seen as a column of dust lofted from the ground. Walking or driving through a dust devil will disrupt its circulation and it may dissipate. Contrast this behavior with that of a tornado, which is not disrupted in the least by anyone crazy enough to venture into one! Dust devils, however, can inflict minor damage. Other, more exotic devils have been documented: A famous video sequence shot in Vermont follows a hay devil. Ash devils have been seen near volcanoes, and fire devils (whirls) in forest fires.

Mountainadoes are vertical vortices associated with intense downslope windstorms that occur during the winter months in the lee of a mountain

6.13 *Steam devil over a golf course in Guerneville, California, at around 10:25–10:30 A.M. on December 10, 1994.*

6.14 *Dust devil somewhere in Nevada on August 7, 1978. Note its similarity in appearance to the waterspout shown in Fig. 2.23.*

range. Windstorms in Boulder, Colorado, that have produced mountaina-does have been documented. One cause of their formation might be bumps in the topography that tilt horizontal vorticity associated with vertical shear of the wind onto the vertical. Counterrotating vertical vortices are produced in the same way that a thunderstorm updraft tilts horizontal vorticity onto the vertical and produces a couplet of vortices (see Fig. 3.4); in a mountainado, the air forced up over the bump plays the same role as the updraft in a thunderstorm.

Yet another small-scale vortex phenomenon has been observed from aircraft on fair-weather days. Looking very much like small waterspouts tilted on their sides, the "wake" vortices appear near or behind a wing (Fig. 6.15).

Common to all but the latter of these small-scale vortices is unstable air near the ground, which causes the air to rise locally. Air converges inward to take its place. If there is any ambient vorticity at the ground, it

6.15 *Horizontal vortex condensation funnel over western Scotland, west of Glasgow, near an aircraft wing, on July 25, 1995.*

will be stretched and amplified to produce an intense vortex. What is a possible source of ambient vorticity near the ground? A variation in vegetation can create horizontal shear from a uniform wind blowing over it because it exerts a horizontal gradient in surface drag. The horizontal vorticity of the wind field may also be the source of vorticity if it is tilted onto the vertical by an updraft.

7

Where We Are Headed

For they have sown the wind,
and they shall
reap the whirlwind . . .
—*Hosea* 8:7

By the mid-1980s computers were big enough and fast enough to simulate the parent vortices of tornadoes. Richard Rotunno and Joe Klemp were the first to do so. Nearly ten years later, meteorologist and modeler Louis Wicker, with the help of more advanced computers, continued this work at an even finer scale. An impressive animation of one of Wicker's simulated tornadoes appears in the 1995 IMAX film *Stormchasers*. Modeler Louis Grasso has also simulated tornado vortices in virtual supercell storms. And recently, modeler Bruce Lee numerically simulated landspout tornadoes, and aerodynamicist Steve Lewellen and collaborators simulated the fine structure of tornadoes. With more advanced computers, the numerical simulation of supercell storms, nonsupercell storms, and the tornadoes each of these spawns will soon be possible.

Richard Anthes and his coworkers demonstrated in 1982 that numerical models could simulate mesoscale features, those on scales of a hundred miles or so, even if the data fed into the model were gathered from the synoptic-scale network (data that can resolve features having scales of a minimum of five hundred miles). Thus, for current mesoscale datasets, although the spacing between data locations might be sixty miles, the models might be able to simulate features having scales of five miles or even smaller. Models using data from the standard observing network (mainly airport sites) can simulate fronts having horizontal scales as small as one tenth of the distance between observing stations.

Will it become possible to start from real data and predict the formation of a real storm, its development into a supercell, and the formation and decay of a tornado within the storm? Some groups are attempting to develop such computer models. They include the Center for the Analysis and Prediction of Storms (CAPS), an NSF-funded science-and-technology center at the University of Oklahoma. With the recent availability of WSR-88D data we can now observe storms on scales relevant to their prediction. The incorporation of these data, Oklahoma mesonet data, and wind-profiler data should help to produce better forecasts by giving the computer a more accurate look at the initial state of the atmosphere. It should also nudge the computer simulation back to the observed state of the atmosphere as the model is pushed into the future. The incorporation of the data from the real world into the model, referred to as *data assimilation,* is now a topic of intense research.

Storm-scale prediction differs from large-scale prediction, which weather forecasters have used for decades. Instead of relying on coarse-resolution models to predict something like a "thirty percent chance of rain in the state this afternoon," as we do now, we soon may use storm-scale models to predict, that, for example, "in four hours, in Cleveland County, in central Oklahoma, there will be a sixty percent chance of hail larger than three quarters of an inch in diameter, and surface wind gusts greater than fifty-five mph."

During the second year of the VORTEX project (1995), the first attempts at predicting where and when thunderstorms will form were made using the CAPS model. We are not yet able to predict the precise location of storms, or to simulate a tornado at the right place and time within a storm. Our limited forecasting ability is reflected in our success rate in accosting tornadoes in the field: On average, we see them only one out of every nine chase days, but on that one day, it is likely that we will see several. The jury is still out as to whether we are using all the data properly and can significantly improve forecasts. Even if we can improve them, a poor or nonexistent road system may hamper the actual chase.

Hurricane-Spawned Tornadoes and Miniature Supercells

Although the central plains of the United States are most commonly thought of as tornado country, tornadoes also are common in hurricanes as they make landfall along the Gulf or southeast coasts. Such tornadoes occur most frequently ahead and to the right of the hurricane's path, where the vertical wind shear is greatest. Unlike the environment of Great Plains supercells, that of hurricane-spawned tornadic storms is characterized by weak buoyancy. Bill McCaul and Morris Weisman, using numerical simulations, have shown how a strong updraft can be forced (from a vertical perturbation-pressure gradient) even when the thermodynamic buoyancy is weak. Relatively weak buoyancy and strong low-level vertical wind shear can produce supercells that are narrower and not so tall as Great Plains supercells. The strongest updraft in the simulated storms tends to be at relatively low levels. Hurricane-spawned tornadoes are rarely, if ever, photographed, probably because copious precipitation cloaks them.

Another aspect of hurricane supercells is that because the air is so moist, the potential for evaporatively cooled downdrafts is very small. Therefore, there is little potential for the generation of horizontal vorticity at the surface associated with density gradients. LP storms also have a low potential for such downdrafts, but this is because there is so little rainfall. Unlike the environment of Great Plains supercells, LP or otherwise, there is little capping inversion at low levels to inhibit storm formation.

Not associated with hurricanes, midget supercells, also called minisupercells, also form in an environment characterized by low buoyancy and strong vertical shear, but near relatively cold pools of air at high levels. A number of them have been documented with the aid of the new WSR-88D Doppler radars. It is a greater challenge for the National Weather Service to issue warnings for these beasts because their smaller mesocyclones are more difficult to detect.

Tornadoes are also spawned in rainbands having no supercell characteristics and far away from the eye of landfalling hurricanes. Little is known about these tornadoes because detailed Doppler-radar analyses of the wind field in them have not yet been documented. They may be similar to landspouts because their vorticity may come from horizontal shear near the ground, which is associated with the rainbands. But unlike landspouts, which occur before rain has fallen, these tornadoes may be hidden in rain.

Hail Detection

Detecting hail by radar is a significant challenge. We cannot tell whether a very intense radar echo is due to backscattering from heavy rain or large hail. Polarized radars are sometimes used to discriminate between hail and drops of water. Polarized sunglasses work on the principle that, under some circumstances, scattered light is polarized. Suppose that sunlight scattered from a certain region of the sky is horizontally polarized. If you look through a piece of polarized glass that admits such light, you will see a bright blue color. On the other hand, if you look at the same region with the glass rotated by ninety degrees, so that only vertically polarized light is admitted, then you will see black or dark blue.

Spherical drops or pieces of ice backscatter electromagnetic radiation at all polarizations; in other words, the oscillations in the electrical and magnetic fields are not restricted to some particular plane. Tiny, round water droplets, of which many clouds are composed, are carried along with the wind with negligible fall velocity. Falling does not deform them and they scatter light equally at all polarizations. Raindrops, on the other hand, are heavy enough to have a terminal fall velocity, and flatten somewhat as they fall. No longer spherical, they do not backscatter radiation equally at all polarizations. Although hailstones may be ellipsoidal and have spikes or other irregularities, most of them tend to be spherical, so they backscatter at all polarizations; if they are irregularly shaped, they spin as they fall and, averaged over time and space, backscatter at all polarizations. We expect backscattered radiation from hail to be less polarized, preferentially in one direction, than backscattered radiation from rain. Radars that can measure the relative intensity of polarized pulses of radiation are called *polarization diversity radars*. Research is now underway using these radars to discriminate between rain and hail, and between different types and sizes of precipitation particles.

But Why Do Storms Form and Why Do Only Some of Them Produce Tornadoes?

We still do not have a good understanding of why storms, tornadic or otherwise, form. We know that air must become buoyant through heating from below, cooling aloft, or both. Moisture must be available to fuel the storms. In many cases, heating at the ground is not sufficient to fire up convection; slow upward motion over a broad region, for example, along fronts or in advance of upper-level disturbance is required to lift air until it becomes buoyant.

The most intense storms seem to be triggered near the boundaries that separate two different air masses, such as fronts (which separate warm from cool air), the dryline (which separates warm, humid air from hot, dry air), and outflow boundaries (which separate evaporatively cooled from warm, humid air). Probably most important is the vertical circulation that tends to be produced at the interface of the two air masses, in particular, the zone of upward motion that contributes to storm formation. In some instances, however, there is no difference in air masses even though the air currents converge along a boundary. It might be that the lift is necessary to cool air to saturation and achieve buoyancy; it might also be that the lift increases moisture above the ground, preventing cumulus towers from losing buoyancy due to the entrainment of dry air. Numerical models, however, can be forced to produce storms that develop into intense virtual supercells even when the storm environment is homogeneous.

We know that even when storms form near a boundary, there often is a Darwinian struggle for survival. Most of the time, neighboring storms interact with each other destructively, and with the exception of large tornado outbreaks, only one or two storms go on to produce tornadoes. What special character must a storm have to produce a tornado? To answer this, we need to know precisely what the differences are between the tornado-producing storm and the non-tornado-producing ones. And to know this, we need to collect and analyze Doppler-radar data documenting the lives and environments of many tornadic and nontornadic storms. As of now, our sample is too small and incomplete. It is possible that the orientation of the storms with respect to the vertical shear is important, because the interaction between neighboring storms depends on storm motion, which in turn depends on the shear. In addition, the way in which storms are positioned with respect to each other depends to some extent on how the boundaries along which they are triggered are oriented.

Another factor that must be considered is low-level stability. If air near the ground can be easily heated enough or lifted just enough to trigger a storm, it is likely that storms will form almost everywhere. Then destructive interference between neighboring cells is very likely. But if low-level stability is high, the likelihood that neighboring cells will be triggered is considerably less.

Bistatic Dual-Doppler Radar Networks

Mobile radar networks and airborne Doppler radars, useful as they are, have one main drawback: They are costly. Josh Wurman demonstrated how to make use of the side scattering off precipitation in convective storms. Because the National Weather Service has already installed WSR-88D Doppler radars across the United States, much less expensive receivers with wide-view antennas can be used to receive side-scattered radiation for a different viewing angle of radar return (from that at the WSR-88D sites) from a storm, anywhere in the country. So, to get dual-Doppler coverage, for example, only one radar system and one receiver are needed, not two complete radar systems. Then one only has to wait for a storm to track through the network. A radar system in which the transmitting and receiving antennas are separated by a considerable distance is called bistatic. Several bistatic networks have recently been set up. It remains to be seen if this approach results in a significant increase in important data sets.

Storm Chasing as a Hobby

In recent years, storm chasers have been featured in numerous newspaper articles, national and international magazines, television programs, and educational and entertainment films and videos, inspiring many adventurers to take up storm chasing for sport. The IMAX film *Stormchasers,* the Hollywood blockbuster film *Twister,* and telecasts by the Weather Channel have without doubt contributed greatly to the upsurge in public interest in storm chasing. A storm chaser's newsletter, *Storm Track,* which was conceived by amateur storm chaser Dave Hoadley in the late 1970s, reaches more than a thousand readers. Television stations all over the globe send their crews to Tornado Alley in the hope of capturing the excitement of a storm chase and the beauty of a severe thunderstorm. Storm-chasing television crews in the Oklahoma City area transmit live video of tornadoes via cellular phone lines or satellite. Pandora's box has been opened, and it is no longer easy to go out into the field and be alone with Nature. You can even book a chase vacation and be guided around Tornado Alley by a seasoned storm-chase veteran.

If you can afford it, flying to an area where storms might be brewing can be rewarding. I recall University of Oklahoma students taking a plane to Dallas, where they rented a car and proceeded to chase nearby storms.

In May 1996 I flew from Denver to Omaha, Nebraska, with a reporter from *People* magazine, where we joined a photographer from Kansas City, rented a Dodge Stratus (named after the least convective of all clouds), and went after a tornadic storm that moved from South Dakota into Minnesota. We ended up, after chasing all afternoon and early evening, in St. Cloud, Minnesota—a place with a rather fitting name.

With the numbers of storm chasers burgeoning, a new idiom has evolved for describing tornadoes. One hears about hoses, spikes, tubes, elephants' trunks, wedges, cylinders, and stovepipes. But to the best of my knowledge, storm chasers never use the expression "twister" to describe a tornado; it's considered gauche!

Storm chasing is not without its risks, although most people assume, incorrectly, that the greatest danger when storm chasing is the tornado itself. On the contrary, it is usually very difficult even for researchers, who are relatively skilled at getting near tornadoes, to get close enough to a tornado to be in danger of being tossed about or hit by flying debris. In any event, professional storm chasers *never* try to get so close to tornadoes that tractors fall in front of them or multiple vortices dance about them, as in the movie *Twister*. (A drive-in movie theater in Canada was hit by a tornado screening *Twister* in 1996, although given the number of locations the movie was shown at, it was bound to happen!) Our relationship with tornadoes is like that between lions and lion tamers: Experience allows us to get what appears to the uninitiated to be dangerously close, but we are not really *that* close. The main dangers of storm chasing include driving on rough rural roads in bad weather, especially when the driver is distracted by events outside the chase vehicle. Lightning is also a significant danger; one never knows when it will strike. Flash floods can strand vehicles, members of the bovine community may block roads, large hail cracks or breaks windshields and destroys side-view mirrors, high winds can overturn vehicles, and the caravans of chasers themselves all converging on the same location present significant hazards. During the spring of 1996, one chaser was struck by lightning and injured in southwestern Oklahoma while filming a storm. A University of Oklahoma undergraduate was killed in 1984 while chasing a storm when the driver lost control trying to avoid a cow that had wandered onto the road.

Tornado Modification and Coping with Tornadoes

The expression "tornado modification" has been used in reference to tornado mitigation, not tornado enhancement. We cannot really hope to attempt to modify tornadoes until we learn precisely why they form. Attempts to modify them in the absence of a hypothesis that can be rigorously tested is like shooting in the dark. If we were to try to prevent the tornado's parent storm from forming, farmland might be deprived of much-needed rain. We might succeed in preventing tornadoes, but the consequences might be to increase the size and frequency of large hail or the intensity of surface straight-line winds (as opposed to winds in tornadoes) or the amount of rain, the latter of which could lead to flooding. In other words, the overall damage and injury from nontornado phenomena might be worse than those caused by tornadoes themselves.

Tornado folklore tells us that tornadoes do not strike towns that lie in valleys and are partially or fully surrounded by hills. There is evidence that hilly terrain can weaken a tornado vortex, but there is also documentation of strong tornadoes riding over high ridges and continuing unabated in their damage paths. On July 21, 1987, what is known as the Teton-Yellowstone tornado in northwest Wyoming whipped over high, mountainous terrain, inflicting extensive damage, uprooting and blowing over forests of trees. Funnel clouds over the Colorado high country are common. Shortly after noon one day in July 1988, as I was leaving a breakfast restaurant in downtown Boulder, Colorado, a street person approached me with the greeting, "Hey man, wanna see a couple of tornadoes? Just walk up the street." He could not possibly have been aware of my interest in tornadoes. But there in the west I could see convective clouds, so I ran up the street to the point where I could see two funnel clouds pendant from the cloud base. Moments earlier, it turned out, one of those funnels had been a landspout tornado just east of the Continental Divide near Gold Hill, at about eight thousand feet above sea level. And in July 1996 a supercell tornado uprooted and destroyed trees in the mountains near Colorado Springs. I was hiking far to the north but was lucky enough to film the storm. It therefore seems that suggestions by some to migitate tornadoes by erecting artificial mounds and ridges near cities would not be effective in warding off tornadoes.

Another way to modify tornadoes might be through cloud seeding. An increased amount of rain in the parent storm might increase the outflow of cold air and cut off the supply of warm, moist air from the storm's updraft. The demise of the updraft could decrease the chances for tornado formation. It has also been suggested that exhaust from a high-flying jet could induce a layer of cirrus clouds, which would cut down on the surface heating—but only during the day. Another technique would be to explode a bomb in the tornado's path, disrupting the vortex; heat would be introduced, which would also have the effect of altering the air flow because of the added buoyancy. Of course, the bombs could probably do at least as much damage as the tornado.

In 1975 physicist J. D. Isaacs and colleagues suggested in the journal *Nature* that "vorticity pollution" from motor traffic, which in North America passes opposing traffic on the right, could account for the relatively high incidence of tornadoes in the United States in the latter half of the twentieth century. Cyclonic whirls could be set up in the zone between the opposing streams of traffic. If this were true, the incidence of tornadoes could be cut simply by decreasing motor traffic, but we find many tornadoes in rural areas with light traffic. Ed Kessler pointed out that as a motor vehicle travels, vortices of opposing rotation form in its wake; if an updraft, which is typically much wider than a motor vehicle, is associated with air converging into the vortices, no net vorticity will be produced. Furthermore, anticyclonic vortices should be produced along the edges of the road. The net amount of this anticyclonic vorticity should be equal, then, to the cyclonic vorticity produced between the opposing streams of traffic.

The best way to cope with tornadoes is to keep abreast of local weather forecasts through radio and television, or by NOAA weather radio. NOAA weather-radio broadcasts near 162 MHz can be heard at many locations on inexpensive weather radios, which can be purchased at electronics stores. If severe weather is forecast for later in the day, you should leave a radio or television on to listen for watches and local warnings. Tornado warnings are issued by the National Weather Service when a reliable observer reports a tornado or when a mesocyclone signature is detected by Doppler radar. In some towns the warning may be accompanied by a siren. Some special weather radios have alarm tones that go off when a warning is issued. Some television stations issue their own tornado warnings based on their own Doppler radar and observer reports. If National Weather Service warnings and unofficial television-based warnings are not identical, the public may be misled or confused. It is probably safest and prudent to assume that any warning is valid. If a tornado is about to hit, you probably don't have time to worry about the source of the warning.

When a warning is issued, you should go to your basement or storm cellar. Running outside to take a picture or shoot a video of the approaching tornado is foolhardy. If you cannot seek shelter underground, go to an interior room, preferably a bathroom. It is remarkable how even in severely damaged houses, interior bathrooms often suffer the least damage. Stay away from windows to avoid flying glass and be sure to protect your head from flying debris. Do not open windows: The pressure inside a building should become equal to the lower pressure outside rapidly enough so that the structure will not explode. Opening windows increases the risk of sending damaging winds through a building and may reduce the precious time you have to seek shelter. Do not get into your car and try to outrun a tornado. You—and other panicked drivers—may not be able to get away before the tornado overtakes you, especially if you're all involved in a traffic jam. Mobile homes, even those that are tied down, are particularly susceptible to damage from tornadoes. (It has been suggested many times to us that we should paint the picture of a mobile home onto the side of our storm-intercept van as tornado bait.) If you are at work or in a building away from home, you should follow instructions for reaching the safest place; if there are no instructions, go to the basement, if there is one, or to an interior room on the ground floor. If you are in the open, try to lie down in a low-lying spot (avoiding areas where flash floods might occur) and protect your head.

The most difficult situation is the approach of a tornado late at night, when most people are sleeping. Then you must rely on a local siren or NOAA weather-radio alarm tone to wake you so that you may seek shelter.

—⟨∘⟩—

We have come a long way since the Thunderstorm Project in our understanding of storms that spawn tornadoes and of tornadoes themselves. We still haven't completely solved the puzzle of why tornadoes form, and we may never be able to predict a tornado with a lead time of more than tens of minutes. Yet our efforts have led to the development of new instruments, such as Doppler radar, that are being used to increase the lead time of warnings (which used to be only a few minutes at best) by as much as thirty minutes. Doppler radar has also led to more accurate warnings, and the provision of wind data that can be used in numerical models to increase our forecasting skills.

The source of rotation in some tornadoes appears to be preexisting vortices near the ground above where convective storms are growing; in others, the source seems to be linked to the mesocyclone. Does the mesocyclone itself descend to the ground and intensify to become the tornado, or does it trigger events near the ground that create tornadoes from other sources of rotation? Can it do both? We hope the answers to these questions are soon forthcoming. Our quest for discovery has not taken away from the respect we have for the awesome power the tornado harbors, nor the thrill of viewing the violent motions in the tornado or the beauty of the storm. We eagerly await the next act in the atmospheric play starring the tornado.

Appendix A

The Dynamic Pressure

LET US ASSUME that the air outside a buoyant bubble does not mix with the air inside. As the bubble accelerates up, air is forced out of the way. Underneath the bubble, air is forced into the space the bubble just left. For this to happen, the pressure immediately above the bubble must be higher than off to the side, and underneath it must be lower than off to its side (Fig. A1). The pressure-gradient forces that accelerate air around the bubble are due to the accelerating upward motion of the bubble itself.

The situation is different for a neutrally buoyant bubble rising at a *constant* speed. In both cases, air moves down relative to the rising bubble, as depicted in Fig. A2, and air is diverted around the bubble. Consistent with flow around the bubble *that doesn't change with time* (because the bubble is not accelerating), the pressure is relatively high on the upwind side (top) *and* on the downwind (bottom) side of the bubble. In addition, the pressure on the right and left sides of the bubble is relatively low.

Think of an air current flowing uniformly over the wing of an airplane. The wing is shaped so that much of the air flows up and over it. What forces the air up? The pressure above the wing must be lower than the pressure below. The upward-directed pressure gradient force is what keeps

163

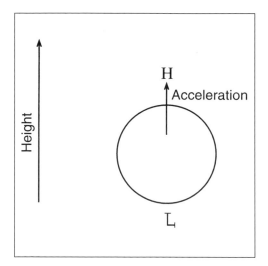

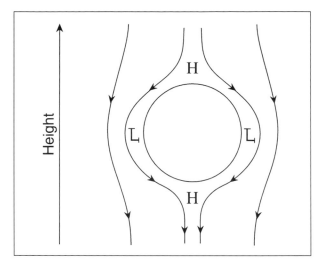

A1   *An idealized illustration of the distribution of the deviation of the pressure from its hydrostatic value about a buoyant bubble (circle) that is accelerating upward. H (L) denotes relatively high (low) pressure.*

A2   *An idealized illustration of the relation between the wind field (depicted by streamlines) in the reference frame of a neutrally buoyant (i.e., having exactly zero buoyancy) bubble (circle) moving upward at a constant speed (with respect to the bubble the wind comes from above) and the deviation of the pressure field from its hydrostatic value. H (L) denotes relatively high (low) pressure.*

the plane in the air. The pressure field is thus associated with the wind field (relative to the wing) and each owes its existence to the other.

I have just described an instantaneous relation between the motion of air and the pressure field. We know that what really happens is that compression (sound) waves travel at high speed, communicating "information" about the bubble or wing to the surrounding air. The anelastic approximation discussed earlier gets rid of the waves by positing the atmosphere to be nearly incompressible; we thus assume that the compression waves travel infinitely quickly, and the air knows instantaneously about the bubble's or wing's motion. The pressure field is that which is necessary to move the air so that its mass is conserved.

We now have a feeling for how the pressure and wind fields are related. Part of the wind field is produced by horizontal hydrostatic pressure gradients, and part is linked to gradients of the dynamic part of the pressure field—not necessarily with any cause and effect. But the pressure-gradient forces associated with the pressure field can *change* the motion field, which would then be associated with another distribution of pressure.

Appendix B

The Effects of Momentum Transport by an Updraft in a Sheared Environment

SUPPOSE A BUOYANT bubble rises, bringing with it lower momentum from below, into an environment of higher wind speed (and higher momentum). At a given instant, as air impinges on the bubble on its upshear side (to its west), the surrounding air is diverted around the sides of the bubble; and as it flows away from the bubble on the downshear side (to its east), the air is augmented from above, below, and the sides. The air that catches up to the bubble slows down as a result of an opposing pressure-gradient force; it speeds up on the opposite side the bubble as a result of a pressure-gradient force from the same direction as its motion. Therefore, the pressure must be relatively high on the upshear side and low on the downshear side (Fig. B1).

Suppose that the intensity of the regions of low and high pressure is greatest in the middle portion of the troposphere, where the combined effects of buoyancy and shear are greatest. At low levels the pressure-gradient force is upward on the downshear side and downward on the upshear side (see Fig. 3.5). Remember that the hydrostatic part of the pressure field is not considered here; it is in balance with the gravity field. It follows then that air will be lifted at low levels on the downshear side and pushed down on the upshear side. If air condenses, becoming buoyant on the downshear

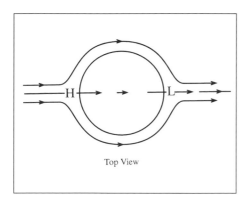

B1 *Relationship between the deviation of pressure from its hydrostatic value and the wind field in a bubble that has risen in an environment of unidirectional vertical wind shear and brought up with it lower momentum from below (denoted by weaker wind vector at the center of the bubble than at the upshear (left) and downshear (right) sides), looking down on the bubble. H (L) denotes relatively high (low) pressure. Some air is diverted around the bubble, while some decelerates as it passes into the bubble and accelerates as it exits the bubble. The deviation of the pressure from its hydrostatic value is relatively high (low) on the upshear (downshear) side.*

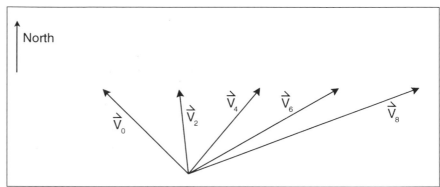

B2 *An example of the variation of wind with height in which the vertical wind shear is westerly at all levels, even though the wind veers from southeasterly to southwesterly with height. The subscript to the right of each wind vector indicates the height above the ground in kilometers.*

side, while the development of buoyant bubbles is suppressed on the upshear side, then the storm will propagate in the downshear direction, which, in this example, is to the east.

Now let's allow the discussion to get more complicated and more realistic. Suppose that the profile of wind shear does not change direction with height, as in Fig. B2. This does not necessarily mean that the wind itself does not change direction with height. For example, we see that the wind shear (i.e., a difference between the wind above and below) is always a wind that blows to the east, while the wind itself veers from southeast to southwest with height.

Suppose now that the vertical wind-shear vector points to the north at low levels, to the east at middle levels, and to the south at high levels (Fig. 3.6). In other words, the vertical wind-shear vector veers, or rotates clockwise with height. At a given level we know from our analysis in Fig. B1 that the pressure is relatively high on the upshear side of a buoyant updraft and low on the downshear side. So pressure to the south of the updraft is relatively high at low levels and low at high levels. There is therefore an upward-directed pressure-gradient force that can trigger new cumulus growth on the south side of the original updraft. Pressure to the north of the updraft is relatively low at low levels and high at high levels. Thus there is a downward directed pressure-gradient force that can suppress cumulus growth on the north side of the updraft. The net effect of all these pressure forces is to make the storm cell propagate to the south (i.e., to the right of the mean vertical shear, which is westerly). The effect is greatest for strong updrafts and strong vertical shear.

If the vertical wind-shear vector backs with height, that is, it rotates counterclockwise with height, then the storm cell will propagate to the north, to the left of the mean vertical shear. In the real atmosphere, the vertical shear, when it is relatively strong, typically veers with height.

Appendix C

Other Resources

Web Sites

Some sites on the World Wide Web well worth visiting are as follows. URLs are current as of July 1998 but subject to change.

Web Site	URL
American Meteorological Society	http://www.ametsoc.org
Colorado State University	http://www.cira.colostate.edu
CNN	http://www.cnn.com
Forecast Systems Laboratory (FSL)	http://www.fsl.noaa.gov
Michigan State University	http://www.wunderground.com
National Center for Atmospheric Research (NCAR)/Research Applications Program (RAP)	http://www.rap.ucar.edu/weather
National Severe Storms Laboratory (NSSL)	http://www.nssl.noaa.gov
National Weather Service	http://iwin.nws.noaa.gov
Ohio State	http://asp1.sbs.ohio-state.edu

Web Site	URL
Oklahoma Weather Center	http://geowww.gcn.ou.edu
Oklahoma Weather Roundup	http://weather.ou.edu/wx/
Purdue University	http://wxp.atms.purdue.edu
Storm Chaser Homepage	http://taiga.geog.niu.edu/chaser.html
Storm Prediction Center (SPC)	http://www.nssl.noaa.gov/~spc
University of Michigan	http://cirrus.sprl.umich.edu/wxnet
University of Wisconsin Space Science and Engineering Center (SSEC)	http://www.ssec.wisc.edu
The Weather Channel	http://www.weather.com
Weatherwise magazine	http://www.heldref.org/ww/ww.html

Organizations

American Meteorological Society
45 Beacon Street
Boston, MA 02108
(Contact the American Meteorological
Society especially if you are interested in
pursuing a career in meteorology.)

National Severe Storms Laboratory
1313 Halley Circle
Norman, OK 73069

National Weather Association
6704 Wolke Court
Montgomery, AL 36116-2134

Texas Severe Storms Association (TESSA)
P.O. Box 122222
Arlington, TX 76012

University Corporation for Atmospheric
Research (UCAR)
P.O. Box 3000
Boulder, CO 80307-3000

The Weather Channel
2600 Cumberland Parkway
Atlanta, GA 30339

Magazines and Newsletters for the Amateur Weather Enthusiast

American Weather Observer
401 Whitney Boulevard
Belvidere, IL 61008-3772

Stormtrack
Tim Marshall, editor
4041 Bordeaux Circle
Flower Mound, TX 75028
(This publication is highly recommended
if you are interested in storm chasing.)

The Weather Bulletin
Texas Severe Storms Association
P.O. Box 122222
Arlington, TX 76012

Weatherwise
Heldref Publications
1319 Eighteenth Street, NW
Washington, DC 20036-1802

References

The following list of references is given in part for its historical value and in part for the more serious reader who wishes to learn more about the technical aspects of topics presented in this book.

Abdullah, A. J., 1966: The "musical" sound emitted by a tornado. *Mon. Wea. Rev.,* **94,** 213–220.

Agee, E. M., 1969: Tornado project activities, Purdue University. *Bull. Amer. Meteor. Soc.,* **50,** 806.

Agee, E. M., 1970: Tornado project activities, Purdue University, Part II. *Bull. Amer. Meteor. Soc.,* **51,** 951.

Agee, E. M., 1971: Tornado project activities, Purdue University, Part III. *Bull. Amer. Meteor. Soc.,* **52,** 575.

Agee, E. M., J. T. Snow, and P. R. Clare, 1976: Multiple vortex features in the tornado cyclone and the occurrence of tornado families. *Mon. Wea. Rev.,* **104,** 552–563.

Anthes, R. A., Y.-H. Kuo, S. G. Benjamin, and Y.-F. Li, 1982: The evolution of the mesoscale environment of severe local storms: Preliminary modeling results. *Mon. Wea. Rev.,* **110,** 1187–1213.

Armijo, L., 1969: A theory for the determination of wind and precipitation velocities with Doppler radars. *J. Atmos. Sci.,* **26,** 570–573.

Atlas, D., 1963: Radar analysis of severe storms. *Meteor. Monograph,* **5,** no. 27, 177–220.

Barcilon, A. I., and P. G. Drazin, 1972: Dust devil formation. *Geophys. Fluid Dyn.,* **4,** 147–158.

Barnes, S. L., 1968: On the source of thunderstorm rotation. *ESSA Technical Memorandum, ERLTM-NSSL 38,* National Severe Storms Laboratory, Norman, OK, 28 pp.

Bates, F. C., 1963: An aerial observation of a tornado and its parent cloud. *Weather,* **18,** 12–18.

Battan, L. J., 1961: *The Nature of Violent Storms.* Doubleday, Garden City, NY, 158 pp.

Bedard, A. J. Jr., and C. Ramzy, 1983: Surface meteorological observations in severe thunderstorms, Part I. Design details of TOTO. *J. Appl. Meteor.,* **22,** 911–918.

Bergen, W. R., 1976: Mountainadoes: A significant contribution to mountain windstorm damage? *Weatherwise,* **29,** 64–69.

Blench, B.J.R., 1963: Luke Howard and his contribution to meteorology. *Weather,* **18,** 83–92.

Bluestein, H., 1979: A mini-tornado in California. *Mon. Wea. Rev.,* **107,** 1227–1229.

Bluestein, H., 1980: The University of Oklahoma Severe Storms Intercept Project—1979. *Bull. Amer. Meteor. Soc.,* **61,** 560–567.

Bluestein, H. B., 1983: Surface meteorological observations in severe thunderstorms. Part II: Field experiments with TOTO. *J. Climate Appl. Meteor.,* **22,** 919–930.

Bluestein, H. B., 1988: Funnel clouds pendant from high-based cumulus clouds. *Weather,* **43,** 220–221.

Bluestein, H. B., 1992: *Synoptic-Dynamic Meteorology in Midlatitudes, Vol. I: Principles of Kinematics and Dynamics.,* Oxford University Press, New York, 431 pp.

Bluestein, H. B., 1993: *Synoptic-Dynamic Meteorology in Midlatitudes, Vol. II: Observations and Theory of Weather Systems,* Oxford University Press, New York, 594 pp.

Bluestein, H. B., 1994: High-based funnel clouds in the Southern Plains. *Mon. Wea. Rev.,* **122,** 2631–2638.

Bluestein, H. B., S. G. Gaddy, D. C. Dowell, A. L. Pazmany, J. C. Galloway, R. E. McIntosh, and H. Stein, 1997: Doppler radar observations of substorm-scale vortices in a supercell. *Mon. Wea. Rev.*, **125**, 1046–1059.

Bluestein, H. B., and J. H. Golden, 1993: A review of tornado observations. In *The Tornado: Its Structure, Dynamics, Prediction, and Hazards.* Geophysical Monographs 79, American Geophysical Union, Washington, DC, 319–352.

Bluestein, H. B., and M. H. Jain, 1985: Formation of mesoscale lines of precipitation: Severe squall lines in Oklahoma during the spring. *J. Atmos. Sci.*, **42**, 1711–1732.

Bluestein, H. B., J. G. LaDue, H. Stein, D. Speheger, and W. P. Unruh, 1993: Doppler radar wind spectra of supercell tornadoes. *Mon. Wea. Rev.*, **121**, 2200–2221.

Bluestein, H. B., E. W. McCaul Jr., G. P. Byrd, and G. Woodall, 1988: Mobile sounding observations of a tornadic storm near the dryline: The Canadian, Texas, storm of 7 May 1986. *Mon. Wea. Rev.*, **116**, 1790–1804.

Bluestein, H. B., and S. Parker, 1993: Modes of isolated, severe convective-storm formation along the dryline. *Mon. Wea. Rev.*, **121**, 1354–1372.

Bluestein, H., and C. Parks, 1983: A synoptic and photographic climatology of low-precipitation severe thunderstorms in the Southern Plains. *Mon. Wea. Rev.*, **111**, 2034–2046.

Bluestein, H. B., A. L. Pazmany, J. C. Galloway, and R. E. McIntosh, 1995: Studies of the substructure of severe convective storms using a mobile 3-mm wavelength Doppler radar. *Bull. Amer. Meteor. Soc.*, **76**, 2155–2169.

Bluestein, H. B., and C. J. Sohl, 1979: Some observations of a splitting severe thunderstorm. *Mon. Wea. Rev.*, **107**, 861–873.

Bluestein, H. B., and W. P. Unruh, 1989: Observations of the wind field in tornadoes, funnel clouds, and wall clouds with a portable Doppler radar. *Bull. Amer. Meteor. Soc.*, **70**, cover and 1514–1525.

Bluestein, H. B., W. P. Unruh, D. C. Dowell, T. A. Hutchinson, T. M. Crawford, A. C. Wood, and H. Stein, 1997: Doppler-radar analysis of the Northfield, Texas, tornado of 25 May 1994. *Mon. Wea. Rev.*, **125**, 212–230.

Bluestein, H. B., and G. R. Woodall, 1990: Doppler-radar analysis of a low-precipitation severe storm. *Mon. Wea. Rev.*, **118**, 1640–1664.

Bohren, C. F., and A. B. Fraser, 1993: Green thunderstorms. *Bull. Amer. Meteor. Soc.*, **74**, 2185–2193.

Brady, R. H., and E. J. Szoke, 1989: A case study of nonmesocyclone tornado development in northeast Colorado. *Mon. Wea. Rev.*, **117**, 843–856.

Braham, R. R. Jr., 1996: The Thunderstorm Project. *Bull. Amer. Meteor. Soc.*, **77**, 1835–1845.

Brandes, E. A., 1977: Gust front evolution and tornado genesis as viewed by Doppler radar. *J. Appl. Meteor.*, **16**, 333–338.

Brandes, E. A., 1978: Mesocyclone evolution and tornadogenesis: Some observations. *Mon. Wea. Rev.*, **106**, 995–1011.

Brandes, E. A., 1981: Finestructure of the Del City-Edmond tornadic mesocirculation. *Mon. Wea. Rev.*, **109**, 635–647.

Brandes, E. A., 1984: Relationships between radar derived thermodynamic variables and tornadogenesis. *Mon. Wea. Rev.*, **112**, 1033–1052.

Brandes, E. A., 1993: Tornadic thunderstorm characteristics determined with Doppler radar. In *The Tornado: Its Structure, Dynamics, Prediction, and Hazards.* Geophysical Monographs 79, American Geophysical, Union, Washington, DC, 143–159.

Bringi, V. N., and A. Hendry, 1990: Technology of polarization diversity radars for meteorology. In *Radar in Meteorology: Battan Memorial and 40th Anniversary Radar Meteorology Conference* (D. Atlas, ed.), American Meteorological Society, Boston, 153–190.

Brooks, E. M., 1949: The tornado cyclone. *Weatherwise*, **2**, 32–33.

Brooks, H. E., C. A. Doswell III, and J. Cooper, 1994: On the environments of tornadic and nontornadic mesocyclones. *Wea. and Forecasting*, **9**, 606–618.

Brooks, H. E., C. A. Doswell III, and R. B. Wilhelmson, 1996: The role of midtropospheric winds in the evolution and maintenance of low-level mesocyclones. *Mon. Wea. Rev.*, **122**, 126–136.

Brooks, H. E., and R. B. Wilhelmson, 1993: Hodograph curvature and updraft intensity in numerically modeled supercells. *J. Atmos. Sci.*, **50**, 1824–1833.

Brown, J. M., and K. R. Knupp, 1980: The Iowa cyclonic-anticyclonic tornado pair and its parent thunderstorm. *Mon. Wea. Rev.*, **108**, 1626–1646.

Brown, R. A., W. C. Bumgarner, K. C. Crawford, and D. Sirmans, 1971: Preliminary Doppler velocity measurements in a developing radar hook echo. *Bull. Amer. Meteor. Soc.*, **52**, 1186–1188.

Brown, R. A., D. W. Burgess, J. K. Carter, L. R. Lemon, and D. Sirmans, 1975: NSSL dual-Doppler radar measurements in tornadic storms: A preview. *Bull. Amer. Meteor. Soc.*, **56**, 524–526.

Brown, R. A., L. R. Lemon, and D. W. Burgess, 1978: Tornado detection by pulsed Doppler radar. *Mon. Wea. Rev.*, **106**, 29–38.

Browning, K. A., 1964: Airflow and precipitation trajectories within severe local storms which travel to the right of the winds. *J. Atmos. Sci.*, **21**, 634–639.

Browning, K. A., 1965: The evolution of tornadic storms. *J. Atmos. Sci.*, **22**, 664–668.

Browning, K. A., 1965: A family outbreak of severe local storms—a comprehensive study of the storms in Oklahoma on 26 May 1963: Part 1. AFCRL-65-695 (1) Special Report no. 32, 346 pp. (Available from Defense Technical Information Center [DTIC], Ft. Belvoir, VA 22060-6218; AD-623787.)

Browning, K. A., 1977: The structure and mechanisms of hailstorms. In *Hail: A Review of Hail Science and Hail Suppression* (G. B. Foote and C. A. Knight, eds.), *Meteorological Monographs*, no. 38, American Meteorological Society, Boston, 1–43.

Browning, K. A., and R. J. Donaldson Jr., 1963: Airflow and structure of a tornadic storm. *J. Atmos. Sci.*, **20**, 533–545.

Browning, K. A., and G. B. Foote, 1976: Airflow and hail growth in supercell storms and some implications for hail suppression. *Quart. J. Roy. Meteor. Soc.*, **102**, 499–533.

Browning, K. A., and F. H. Ludlam, 1962: Airflow in convective storms. *Quart. J. Roy. Meteor. Soc.*, **88**, 117–135.

Burgess, D. W., 1976: Anticyclonic tornado. *Weatherwise*, **29**, cover and 167.

Burgess, D. W., and R. P. Davies-Jones, 1979: Unusual tornadic storms in eastern Oklahoma on 5 December 1975. *Mon. Wea. Rev.*, **107**, 451–457.

Burgess, D. W., R. J. Donaldson Jr., and P. R. Desrochers, 1993: Tornado detection warning by radar. In *The Tornado: Its Structure, Dynamics, Prediction, and Hazards.* Geophysical Monographs 79, American Geophysical Union, Washington, DC, 203–221.

Burgess, D. W., and L. R. Lemon, 1990: Severe thunderstorm detection by radar. In *Radar in Meteorology* (D. Atlas, ed.), American Meteorological Society, Boston, 619–647.

Busk, H. G., 1927: Land waterspouts. *Meteor. Magazine*, 289–291.

Byers, H. R., and R. R. Braham, 1949: *The Thunderstorm.* U.S. Government Printing Office, Washington, DC, 287 pp.

Carbone, R. E., 1983: A severe frontal rainband. Part II: Tornado parent vortex circulation. *J. Atmos. Sci.*, **40**, 2639–2654.

Carlson, T. N., S. G. Benjamin, G. S. Forbes, and Y.-F. Li, 1983: Elevated mixed layers in the regional severe storm environment: Conceptual model and case studies. *Mon. Wea. Rev.*, **111**, 1453–1473.

Carlson, T. N., and F. H. Ludlam, 1968: Conditions for the occurrence of severe local storms. *Tellus*, **20**, 203–226.

Charba, J., 1974: Application of gravity current model to analysis of squall-line gust front. *Mon. Wea. Rev.*, **102**, 140–156.

Charba, J., and Y. Sasaki, 1971: Structure and movement of severe thunderstorms of 3 April 1964 as revealed from radar and surface mesonetwork data analysis. *J. Meteor. Soc. Japan*, **49**, 191–214.

Chisholm, A. J., and J. H. Renick, 1972: The kinematics of multicell and supercell Alberta hailstorms. In *Alberta Hail Studies*, Report no. 72-2, Research Council of Alberta Hail Studies, 24–31.

Church, C. R., and J. T. Snow, 1993: Laboratory models of tornadoes. In *The Tornado: Its Structure, Dynamics, Prediction, and Hazards*. Geophysical Monographs 79, American Geophysical Union, Washington, DC, 277–295.

Church, C. R., J. T. Snow, and E. M. Agee, 1977: Tornado vortex simulation at Purdue University. *Bull. Amer. Meteor. Soc.*, **58**, 900–908.

Church, C. R., J. T. Snow, G. L. Baker, and E. M. Agee, 1979: Characteristics of tornado-like vortices as a function of swirl ratio: A laboratory investigation. *J. Atmos. Sci.*, **36**, 1755–1776.

Church, C. R., J. T. Snow, and J. Dessens, 1980: Intense atmospheric vortices associated with a 1000 MW fire. *Bull. Amer. Meteor. Soc.*, **61**, 682–694.

Clary, M., 1986: Chasing tornadoes. *Weatherwise*, **39**, 136–145.

Cooley, J. R., 1968: Cold air funnel clouds. *Mon. Wea. Rev.*, **106**, 1368–1372.

Daily Camera, 1996: Tornado survivors praise Allah. Boulder, CO, 16 May, 4A.

Daily Camera, 1997: Twister drops men, house in nearby lake, 18 Aug, 5A.

Davies-Jones, R. P., 1979: Dual-Doppler radar coverage area as a function of measurement accuracy and spatial resolution. *J. Appl. Meteor.*, **18**, 1229–1233.

Davies-Jones, R. P., 1984: Streamwise vorticity: The origin of updraft rotation in supercell storms. *J. Atmos. Sci.*, **41**, 2991–3006.

Davies-Jones, R. P., 1986: Tornado dynamics. In *Thunderstorm Morphology and Dynamics* (E. Kessler, ed.), University of Oklahoma Press, Norman, 197–236.

Davies-Jones, R. P., 1995: Tornadoes. *Sci. American*, **273**, 48–59.

Davies-Jones, R. P., D. W. Burgess, and L. R. Lemon, 1976: An atypical tornado-producing cumulonimbus. *Weather*, **31**, 337–347.

Davies-Jones, R. P., D. W. Burgess, L. R. Lemon, and D. Purcell, 1978: Interpretation of surface marks and debris patterns from the 24 May 1973 Union City, Oklahoma, tornado. *Mon. Wea. Rev.*, **106**, 12–21.

Davies-Jones, R. P., and J. H. Golden, 1975: On the relation of electrical activity to tornadoes. *J. Geophys. Res.*, **80**, 1614–1616.

Davies-Jones, R. P., and J. H. Henderson, 1975: Updraft properties deduced statistically from rawin soundings. *Pure Appl. Geophys.*, **113**, 787–801.

Davies-Jones, R. P., and E. Kessler, 1974: Tornadoes. In *Weather and Climate Modification* (W. N. Hess, ed.), John Wiley and Sons, New York, 552–595.

Day, J. A., and F. H. Ludlam, 1972: Luke Howard and his clouds: A contribution to the early history of cloud physics. *Weather*, **27**, 448–461.

Dennis, A. S., C. A. Schock, and A. Koscielski, 1970: Characteristics of hailstorms of western South Dakota. *J. Appl. Meteor.*, **9**, 127–135.

Dessens, J., 1972: Influence of ground roughness on tornadoes: A laboratory simulation. *J. Appl. Meteor.*, **11**, 72–75.

Donaldson, R. J. Jr., 1970: Vortex signature recognition by a Doppler radar. *J. Appl. Meteor.*, **9**, 666–670.

Donaldson, R. J., 1990: Foundations of severe storm detection by radar. In *Radar in Meteorology: Battan Memorial and 40th Anniversary Radar Meteorology Conference* (D. Atlas, ed.), American Meteorological Society, Boston, 115–121.

Doswell, C. A., and D. W. Burgess, 1988: On some issues of U. S. tornado climatology. *Mon. Wea. Rev.*, **116**, 495–501.

Doswell, C. A. III, and D. W. Burgess, 1993: Tornadoes and tor-nadic storms: A review of conceptual models. In *The Tornado: Its Structure, Dynamics, Prediction, and Hazards*. Geophysical Monographs 79, American Geophysical Union, Washington, DC, 161–172.

Douglas, R. H., 1990: The Stormy Weather Group (Canada). In *Radar in Meteorology: Battan Memorial and 40th Anniversary Radar Meteorology Conference* (D. Atlas, ed.), American Meteorological Society, Boston, 61–68.

Doviak, R. J., P. S. Ray, R. G. Strauch, and L. J. Miller, 1976: Error estimation in wind fields derived from dual-Doppler radar measurements. *J. Appl. Meteor.*, **15**, 868–878.

Dowell, D. C., H. B. Bluestein, and D. P. Jorgensen, 1997: Airborne Doppler radar analysis of supercells during COPS-91. *Mon. Wea. Rev.*, **125**, 365–383.

Dowell, D. C., and H. B. Bluestein, 1997: The Arcadia, Oklahoma, storm of 17 May 1981: Analysis of a supercell during tornadogenesis. *Mon. Wea. Rev.* **125**, 2562–2582.

Elsom, D. M., 1987: Tornadoes in the Soviet Union and the delimitation of regions experiencing devastating tornadoes. *J. Meteor.* (U.K.), **12**, 185–191.

Emanuel, K. A., 1981: A similarity theory for unsaturated downdrafts within clouds. *J. Atmos. Sci.*, **38**, 1541–1557.

Emanuel, K. A., 1994: *Atmospheric Convection*. Oxford University Press, New York, 567 pp.

Eskridge, R. E., and P. Das, 1976: Effect of a precipitation-driven downdraft on a rotating wind field: A possible trigger mechanism for tornadoes? *J. Atmos. Sci.*, **33**, 70–84.

Fankhauser, J. C., 1971: Thunderstorm-environment interactions determined from aircraft and radar observations. *Mon. Wea. Rev.*, **99**, 171–192.

Fawbush, E. J., and R. C. Miller, 1954: The types of airmasses in which North American tornadoes form. *Bull. Amer. Meteor. Soc.*, **35**, 154–165.

Fawbush, E. J., R. C. Miller, and L. G. Starrett, 1951: An empirical method of forecasting tornado development. *Bull. Amer. Meteor. Soc.*, **32**, 1–9.

Felknor, P. S., 1992: *The Tri-State Tornado*. Iowa State University Press, Ames, 131 pp.

Fiedler, B. H., 1996: The sonic speed limit of tornadoes. *Preprints, 18th Conference on Severe Local Storms*, American Meteorological Society, Boston, 385–386.

Fiedler, B. H., and R. Rotunno, 1986: A theory for the maximum windspeeds in tornado-like vortices. *J. Atmos. Sci.*, **43**, 2328–2340.

Fletcher, J. O., 1990: Early developments of weather radar during World War II. In *Radar in Meteorology: Battan Memorial and 40th Anniversary Radar Meteorology Conference* (D. Atlas, ed.), American Meteorological Society, Boston, 3–6.

Flora, S. D., 1953: *Tornadoes of the United States*. University of Oklahoma Press, Norman, 194 pp.

Foote, G. B., and P. S. duToit, 1969: Terminal velocity of raindrops aloft. *J. Appl. Meteor.*, **8**, 249–253.

Fujita, T. T., 1960: A detailed analysis of the Fargo tornadoes of June 20, 1957. U.S. Weather Bureau Research Paper no. 42, 67 pp.

Fujita, T. T., 1960: Mother cloud of the Fargo tornadoes of 20 June, 1957. *Cumulus Dynamics* (C. Anderson, ed.), Pergamon Press, New York, 175–177.

Fujita, T. T., 1970: Lubbock tornadoes: A study of suction spots. *Weatherwise*, **23**, 160–173.

Fujita, T. T., 1973: Tornadoes around the world. *Weatherwise*, 26, 56–62, 78–83.

Fujita, T. T., 1974: Jumbo tornado outbreak of 3 April 1974. *Weatherwise*, 27, 116–126.

Fujita, T. T., 1977: Anticyclonic tornadoes. *Weatherwise*, 30, 51–64.

Fujita, T. T., 1981: Tornadoes and downbursts in the context of generalized planetary scales. *J. Atmos. Sci.*, 38, 1511–1534.

Fujita, T. T., 1985: *The Downburst: Microburst and Macroburst.* Satellite and Mesometeorology Research Project, Department of Geophysical Sciences, University of Chicago, 122 pp.

Fujita, T. T., 1989: The Teton-Yellowstone tornado of 21 July 1987. *Mon. Wea. Rev.*, 117, 1913–1940.

Fujita, T. T., D. L. Bradbury, and C. F. Van Thullenar, 1970: Palm Sunday tornadoes of April 11, 1965. *Mon. Wea. Rev.*, 29, 26–69.

Fujita, T. T., and B. E. Smith, 1993: Aerial survey and photography of tornado and microburst damage. In *The Tornado: Its Structure, Dynamics, Prediction, and Hazards.* Geophysical Monographs 79, American Geophysical Union, Washington, DC, 479–493.

Gal-Chen, T., 1978: A method for the initialization of the anelastic equations: Implications for matching models with observations. *Mon. Wea. Rev.*, 106, 587–606.

Gall, R. L., 1985: Linear dynamics of the multiple-vortex phenomenon in tornadoes. *J. Atmos. Sci.*, 42, 761–772.

Gallagher, F. W. III, W. H. Beasley, and C. F. Bobren, 1996: Green thunderstorms observed *Bull. Amer. Meteor. Soc.*, 77, 2889–2897.

Galway, J. G., 1977: Some climatological aspects of tornado outbreaks. *Mon. Wea. Rev.*, 105, 477–484.

Galway, J. G., 1992: Early severe thunderstorm forecasting and research by the United States Weather Bureau. *Wea. and Forecasting*, 7, 564–587.

Gentry, R. C., 1983: Genesis of tornadoes associated with hurricanes. *Mon. Wea. Rev.*, 111, 1793–1805.

Glaser, A. H., 1960: An observational deduction of the structure of a tornado vortex. In *Cumulus Dynamics* (C. Anderson, ed.), Pergamon Press, New York, 157–166.

Golden, J. H., 1968: Waterspouts at Lower Matecumbe Key, Florida, September 2, 1967. *Weather*, 23, 103–114.

Golden, J. H., 1971: Waterspouts and tornadoes over south Florida. *Mon. Wea. Rev.*, 99, 146–154.

Golden, J. H., 1974: The life of Florida Keys waterspouts, I. *J. Appl. Meteor.*, 13, 676–692.

Golden, J. H., and B. J. Morgan, 1972: The NSSL-Notre Dame Tornado Intercept Program, spring 1972. *Bull. Amer. Meteor. Soc.*, 53, 1178–1180.

Golden, J. H., and D. Purcell, 1977: Photogrammetric velocities for the Great Bend, Kansas, tornado of 30 August 1974: Accelerations and asymmetries. *Mon. Wea. Rev.*, 105, 485–492.

Golden, J. H., and D. Purcell, 1978: Life cycle of the Union City, Oklahoma, tornado and comparison with waterspouts. *Mon. Wea. Rev.*, 106, 3–11.

Grazulis, T. P., 1993: *Significant Tornadoes, 1680–1991.* Environmental Films, St. Johnsbury, VT, 1326 pp.

Hane, C. E., 1973: The squall line thunderstorm: Numerical experimentation. *J. Atmos. Sci.*, 32, 1672–1690.

Hane, C. E., and B. C. Scott, 1978: Temperature and pressure perturbations within convective clouds derived from detailed air motion information: Preliminary testing. *Mon. Wea. Rev.*, 106, 654–661.

Heymsfield, G. M., 1978: Kinematic and dynamic aspects of the Harrah tornadic storm analyzed from dual-Doppler radar data. *Mon. Wea. Rev.*, 106, 233–254.

Hildebrand, P. H., and R. K. Moore, 1990: Meteorological radar observations from mobile platforms. In *Radar in Meteorology: Battan Memorial and 40th Anniversary Radar Meteorology Conference* (D. Atlas, ed.), American Meteorological Society, Boston, 287–314.

Hoecker, W. H., 1960: Wind speed and airflow patterns in the Dallas tornado of April 2, 1957. *Mon. Wea. Rev.*, 88, 167–180.

Houze, R. A. Jr., 1993: *Cloud Dynamics.* Academic Press, San Diego, 573 pp.

Houze, R. A. Jr., B. F. Smull, and P. Dodge, 1990: Mesoscale organization of springtime rainstorms in Oklahoma. *Mon. Wea. Rev.*, 118, 613–654.

Howells, P. A. C., R. Rotunno, and R. K. Smith, 1988: A comparative study of atmospheric and laboratory-analogue numerical tornado-vortex models. *Quart. J. Roy. Meteor. Soc.*, 114, 801–822.

Idso, S. B., 1974: Tornado or dust devil: The enigma of desert whirlwinds. *Amer. Sci.*, 62, 530–541.

Isaacs, J. D., J. W. Stork, D. B. Goldstein, and G. L. Wick, 1975: Effect of vorticity pollution by motor vehicles on tornadoes. *Nature*, 253, 254–255.

Jahn, D. E., and K. K. Droegemeier, 1996: Simulation of convective storms in environments with independently varying bulk Richardson number shear and storm-relative helicity. In *Preprints, 18th Conference on Severe Local Storms*, American Meteorological Society, Boston, 230–234.

Johns, R. H., and C. A. Doswell III, 1992: Severe local storms forecasting. *Wea. and Forecasting*, 7, 588–612.

Johns, R. H., and W. D. Hirt, 1987: Derechos: Widespread convectively induced windstorms. *Weat. and Forecasting*, 2, 32–49.

Johnson, B. C., 1983: The heat burst of 29 May 1976. *Mon. Wea. Rev.*, 111, 1776–1792.

Jones, H. L., 1951: A sferic method of tornado identification and tracking. *Bull. Amer. Meteor. Soc.*, 32, 380–385.

Jorgensen, D. P., and B. F. Smull, 1993: Mesovortex circulations seen by airborne Doppler radar within a bow-echo mesoscale convective system. *Bull. Amer. Meteor. Soc.*, 74, 2146–2157.

Justice, A. A., 1930: Seeing the inside of a tornado. *Mon. Wea. Rev.*, 58, 204–206.

Kangieser, P. C., 1954: A physical explanation of the hollow structure of waterspout tubes. *Mon. Wea. Rev.*, 82, 147–152.

Keller, D., and B. Vonnegut, 1976: Wind speeds required to drive straws and splinters into wood. *J. Appl. Meteor.*, 15, 899–901.

Kelly, D. L., J. T. Schaefer, R. P. McNulty, C. A. Doswell III, and R. F. Abbey Jr., 1978: An augmented tornado climatology. *Mon. Wea. Rev.*, 106, 1172–1183.

Kessler, E., 1969: *On the Distribution and Continuity of Water Substance in Atmospheric Circulations.* Meteor. Mono. 10, American Meteorological Society, Boston, 84 pp.

Kessler, E., 1970: Tornadoes. *Bull. Amer. Meteor. Soc.*, 51, 926–936.

Kessler, E., 1990: Radar meteorology at the National Severe Storms Laboratory, 1964–1986. In *Radar in Meteorology: Battan Memorial and 40th Anniversary Radar Meteorology Conference* (D. Atlas, ed.), American Meteorological Society, Boston, 44–53.

Klemp, J. B., 1987: Dynamics of tornadic thunderstorms. *Ann. Rev. Fluid Mech.*, 19, 369–402.

Klemp, J. B., and R. Rotunno, 1983: A study of the tornadic region within a supercell thunderstorm. *J. Atmos. Sci.*, 40, 359–377.

Klemp, J. B., and R. B. Wilhelmson, 1978: The simulation of three-dimensional convective storm dynamics. *J. Atmos. Sci.*, 35, 1070–1096.

Klemp, J. B., R. B. Wilhelmson, and P. S. Ray, 1981: Observed and numerically simulated structure of a mature supercell thunderstorm. *J. Atmos. Sci.*, 38, 1558–1580.

Koscielski, A., 1967: The Black Hills tornado of 23 June 1966 in South Dakota. *Weatherwise*, 20, 272–274.

Kraus, M. J., 1973: Doppler radar observations of the Brookline, Massachusetts, tornado of 9 August 1972. *Bull. Amer. Meteor. Soc.*, **54**, 519–524.

Lee, B. D., and R. B. Wilhelmson, 1997: The numerical simulation of nonsupercell tornadogenesis. Part II: Evolution of a family of tornadoes along a weak outflow boundary. *J. Atmos. Sci.*, **54**, 2387–2415.

Lee, J. T., D. S. Zrnic, R. P. Davies-Jones, and J. H. Golden, 1981: Summary of AEC-ERDA-NRC supported research at NSSL, 1973–1979. *NOAA Technical Memorandum*, ERL NSSL-90, Norman, OK, 93 pp.

Lemon, L. R., D. W. Burgess, and R. A. Brown, 1978: Tornadic storm airflow and morphology derived from single-Doppler radar measurements. *Mon. Wea. Rev.*, **106**, 48–61.

Lemon, L. R., and C. A. Doswell II, 1979: Severe thunderstorm evolution and mesocyclone structure as related to tornadogenesis. *Mon. Wea. Rev.*, **107**, 1184–1197.

Leslie, L. M., 1971: The development of concentrated vortices: A numerical study. *J. Fluid Mech.*, **48**, 1–21.

Leverson, V. H., P. C. Sinclair, and J. H. Golden, 1977: Waterspout wind, temperature, and pressure structure deduced from aircraft measurements. *Mon. Wea. Rev.*, **105**, 725–733.

Lewellen, W. S., 1990: Tornado vortex theory. In *The Tornado: Its Structure, Dynamics, Prediction, and Hazards.* Geophysical Monographs 79, American Geophysical Union, Washington, DC, 19–39.

Lewellen, W. S., D. C. Lewellen, and R. I. Sykes, 1997: Large-eddy simulation of a tornado's interaction with the surface. *J. Atmos. Sci.*, **54**, 581–605.

Lhermitte, R. M., 1964: Doppler radars as severe storm sensors. *Bull. Amer. Meteor. Soc.*, **45**, 587–596.

Lhermitte, R. M., 1966: Application of pulse Doppler radar techniques in meteorology. *Bull. Amer. Meteor. Soc.*, **47**, 707–711.

Lilly, D. K., 1986: The structure, energetics and propagation of rotating convective storms. Part II: Helicity and storm stabilization. *J. Atmos. Sci.*, **43**, 126–140.

Long, R. R., 1956: Sources and sinks at the axis of a rotating liquid. *Quart. J. Mech. Appl. Math.*, **9**, 385–393.

Lorenz, D., and M. Miller, 1991: *Das 3-D Wolken Buch.* Witting Fachbuch, Huckelhoven, Germany, 247 pp.

Lucretius, C. T., circa 60 B.C.E (1950): *De Rerum Natura*, Metrical translation by W. E. Leonard, E. P. Dutton, New York, 260.

Ludlam, F. H., 1963: Severe local storms: A review. *Meteorological Monographs*, **5**, no. 27, American Meteorological Society, Boston, 1–30.

Ludlum, D. M., 1970: *Early American Tornadoes. 1586–1870.* American Meteorological Society, Boston, 219 pp.

Ludlum, D. M., 1971: The "new champ" hailstone. *Weatherwise*, **24**, 151.

MacGorman, D. R., 1993: Lightning in tornadic storms: A review. In *The Tornado: Its Structure, Dynamics, Prediction, and Hazards.* Geophysical. Monographs 79, American Geophysical Union, Washington, DC, 173–182.

MacGorman, D. R., D. W. Burgess, V. Mazur, W. D. Rust, W. L. Taylor, and B. C. Johnson, 1989: Lightning rates relative to tornadic storm evolution on 22 May 1981. *J. Atmos. Sci.*, **46**, 221–250.

MacGorman, D. R., and K. E. Nielsen, 1991: Cloud-to-ground lightning in a tornadic storm on 8 May 1986. *Mon. Wea. Rev.*, **119**, 1557–1574.

Maddox, R. A., 1976: An evaluation of tornado proximity, wind and stability data. *Mon. Wea. Rev.*, **104**, 133–142.

Maddox, R. A., 1980: Meoscale convective complexes. *Bull. Amer. Meteor. Soc.*, **61**, 1374–1387.

Malkin, W., and J. G. Galway, 1953: Tornadoes associated with hurricanes. *Mon. Wea. Rev.*, **81**, 299–303.

Marshall, T., 1983: Chasing tornadoes. *Weatherwise*, **36**, 184–187.

Marshall, T., 1995: *Storm Talk.* Available from Tim Marshall (see Appendix C), 223 pp.

Marwitz, J. D., 1972: The structure and motion of severe hailstorms. Part I: Supercell storms. *J. Appl. Meteor.*, **11**, 166–179.

Marwitz, J. D., 1972: The structure and motion of severe hailstorms. Part II: Multicell storms. *J. Appl. Meteor.*, **11**, 180–188.

Marwitz, J. D., 1972: The structure and motion of severe hailstorms. Part III: Severely sheared storms. *J. Appl. Meteor.*, **11**, 189–201.

McCarthy, J., G. M. Heymsfield, and S. P. Nelson, 1974: Experiment to deduce tornado cyclone inflow characteristics using chaff and NSSL dual Doppler radars. *Bull. Amer. Meteor. Soc.*, **55**, 1130–1131.

McCaul, E. W. Jr., 1987: Observations of the Hurricane "Danny" tornado outbreak of 16 August 1985. *Mon. Wea. Rev.*, **115**, 1206–1223.

McCaul, E. W. Jr., 1991: Buoyancy and shear characteristics of hurricane tornado environments. *Mon. Wea. Rev.*, **119**, 1954–1978.

McCaul, E. W. Jr., H. B. Bluestein, and R. J. Doviak, 1987: Airborne Doppler lidar observations of convective phenomena in Oklahoma. *J. Atmos. Oceanic Technol.*, **4**, 479–497.

McCaul, E. W. Jr., and M. L. Weisman, 1996: Simulations of shallow supercell storms in landfalling hurricane environments. *Mon. Wea. Rev.*, **124**, 408–429.

Metcalf, J. I., and K. M. Glover, 1990: A history of weather radar research in the U.S. Air Force. In *Radar in Meteorology: Battan Memorial and 40th Anniversary Radar Meteorology Conference* (D. Atlas, ed.), American Meteorological Society, Boston, 32–43.

Miller, R. C., 1959: Tornado-producing synoptic patterns. *Bull. Amer. Meteor. Soc.*, **40**, 465–472.

Miller, R. C., 1972: *Notes on Analysis and Severe-storm Forecasting Procedures of the Air Force Global Weather Central*, Tech. Rep. 200 (rev.), Air Weather Service, Scott Air Force Base, IL, 181 pp.

Moller, A., C. Doswell, J. McGinley, S. Tegtmeier, and R. Zipser, 1974: Field observations of the Union City tornado in Oklahoma. *Weatherwise*, **27**, 68–77.

Morton, B. R., G. I. Taylor, and J. S. Turner, 1956: Turbulent gravitational convection from maintained and instantaneous sources. *Proc. Royal Soc. London*, **A234**, 1–23.

Nelson, S. P., and S. K. Young, 1979: Characteristics of Oklahoma hailfalls and hailstorms. *J. Appl. Meteor.*, **18**, 339–347.

Newton, C. W., 1963: Dynamics of severe convective storms. *Meteor. Monograph*, **5**, no. 27, Amer. Meteor. Soc., Boston, 33–58.

Newton, C. W., and S. Katz, 1958: Movement of large convective rainstorms in relation to winds aloft. *Bull. Amer. Meteor. Soc.*, **39**, 129–136.

Newton, C. W., and H. R. Newton, 1959: Dynamical interactions between large convective clouds and environment with vertical shear. *J. Meteor.*, **16**, 483–496.

New York Times, 1996: Death toll in Bangladesh tornado rises to at least 443. 15 May, A3.

NOAA, 1980: *Storm Data.* **22**, National Climatic Center, Asheville, NC, 15.

Novlan, D. J., and W. M. Gray, 1974: Hurricane spawned tornadoes. *Mon. Wea. Rev.*, **102**, 476–488.

Ogura, Y., 1963: The evolution of a moist convective element in a shallow, conditionally unstable atmosphere: A numerical calculation. *J. Atmos. Sci.*, **20**, 407–424.

Ogura, Y., and N. A. Phillips, 1962: Scale analysis of deep and shallow convection in the atmosphere. *J. Atmos. Sci.*, **19**, 173–179.

Pauley, R. L., C. R. Church, and J. T. Snow, 1982: Measurements of maximum surface pressure deficits in modeled atmospheric vortices. *J. Atmos. Sci.*, **39**, 369–377.

Pauley, R. L., and J. T. Snow, 1988: On the kinematics and dynamics of the 18 July 1986 Minneapolis tornado. *Mon. Wea. Rev.*, 116, 2731–2736.

Pearson, A. D., and A. F. Sadowski, 1965: Hurricane-induced tornadoes and their distribution. *Mon. Wea. Rev.*, 93, 461–464.

Penn, S., C. Pierce, and J. K. McGuire, 1955: The squall line and Massachusetts tornadoes of 9 June 1953. *Bull. Amer. Meteor. Soc.*, 36, 109–122.

Peterson, R. E., 1979: Horizontal funnels—a historical note. *Bull. Amer. Meteor. Soc.*, 60, 795.

Proubert-Jones, J. R., 1990: A history of radar meteorology in the United Kingdom. In *Radar in Meteorology: Battan Memorial and 40th Anniversary Radar Meteorology Conference* (D. Atlas, ed.), American Meteorological Society, Boston, 54–60.

Rasmussen, E. N., J. M. Straka, R. Davies-Jones, C. A. Doswell III, F. H. Carr, M. D. Eilts, and D. R. MacGorman, 1994: Verification of the origins of rotation in tornadoes experiment: VORTEX. *Bull. Amer. Meteor. Soc.*, 75, 995–1006.

Ray, P., 1990: Convective dynamics. In *Radar in Meteorology: Battan Memorial and 40th Anniversary Radar Meteorology Conference* (D. Atlas, ed.), American Meteorological Society, Boston, 348–390.

Ray, P. S., R. J. Doviak, G. B. Walker, D. Sirmans, J. Carter, and B. Bumgarner, 1975: Dual-Doppler observation of a tornadic storm. *J. Appl. Meteor.*, 14, 1521–1530.

Ray, P. S., B. C. Johnson, K. W. Johnson, J. S. Bradberry, J. J. Stephens, K. K. Wagner, R. B. Wilhelmson, and J. B. Klemp, 1981: The morphology of several tornadic storms on 20 May 1977. *J. Atmos. Sci.*, 38, 1643–1663.

Rhea, J. O., 1966: A study of thunderstorm formation along dry lines. *J. Appl. Meteor.*, 5, 58–63.

Rinehart, R. E., 1979: Internal storm motions from a single non-Doppler weather radar. NCAR/TN-146+STR, NCAR, Boulder, CO.

Roberts, R. D., and J. W. Wilson, 1995: The genesis of three nonsupercell tornadoes observed with dual-Doppler radar. *Mon. Wea. Rev.*, 123, 3408–3436.

Rosenfeld, J., 1996: Spin Doctor. *Weatherwise*, 49, 19–25.

Rossmann, F. O., 1960: Some prospects for intercepting tornadoes and saving buildings from explosive destruction. In *Cumulus Dynamics* (C. Anderson, ed.), Pergamon Press, New York, 210–211.

Rotunno, R., 1977: Numerical simulation of a laboratory vortex. *J. Atmos. Sci.*, 34, 1942–1956.

Rotunno, R., 1979: A study in tornado-like vortex dynamics. *J. Atmos. Sci.*, 36, 140–155.

Rotunno, R. 1981: On the evolution of thunderstorm rotation. *Mon. Wea. Rev.*, 109, 577–586.

Rotunno, R., 1993: Supercell thunderstorm modeling and theory. In *The Tornado: Its Structure, Dynamics, Prediction, and Hazards.* Geophysical Monographs 79, American Geophysical Union, Washington, DC. 57–73.

Rotunno, R., and J. B. Klemp, 1982: The influence of the shear-induced pressure gradient on thunderstorm motion. *Mon. Wea. Rev.*, 110, 136–151.

Rotunno, R., and J. B. Klemp, 1985: On the rotation and propagation of simulated supercell thunderstorms. *J. Atmos. Sci.*, 42, 271–292.

Rotunno, R., J. B. Klemp, and M. L. Weisman, 1988: A theory for strong, long-lived squall lines. *J. Atmos. Sci.*, 45, 463–485.

Rust, W. D., R. Davies-Jones, D. W. Burgess, R. A. Maddox, L. C. Showell, T. C. Marshall, and D. K. Lauritsen, 1990: Testing of a mobile version of a cross-chain LORAN atmospheric sounding system (M-CLASS). *Bull. Amer. Meteor. Soc.*, 71, 173–180.

Ryan, J. A., and J. J. Carroll, 1970: Dust devil wind velocities: Mature stage. *J. Geophys. Res*, 75, 531–541.

Schaefer, J. T., 1974: The life cycle of the dryline. *J. Appl. Meteor.*, 13, 444–449.

Schaefer, J. T., 1987: Severe thunderstorm forecasting: A historical perspective. *Wea. and Forecasting*, 1, 164–189.

Schaefer, J. T., D. L. Kelly, and R. F. Abbey, 1986: A minimum assumption tornado-hazard probability model. *J. Climate Appl. Meteor.*, 25, 1934–1945.

Schaefer, J. T., D. L. Kelly, C. A. Doswell III, J. G. Galway, R. J. Williams, R. P. McNulty, L. R. Lemon, and B. D. Lambert, 1980: Tornadoes—when, where, and how often? *Weatherwise*, 33, 52–59.

Schaefer, J. T., and R. L. Livingston, 1988: The typical structure of tornado proximity soundings. *J. Geophys. Res. (Atmospheres)*, 93, D5, 5351–5364.

Schlesinger, R. E., 1973: A numerical model of deep moist convection. Part 1. *J. Atmos. Sci.*, 30, 835–856.

Schlesinger, R. E., 1975: A three-dimensional numerical model of an isolated deep convective cloud: Preliminary results. *J. Atmos. Sci.*, 32, 934–957.

Schlesinger, R. E., 1978: A three-dimensional numerical model of an isolated thunderstorm: Part I: Comparative experiments for variable ambient wind shear. *J. Atmos. Sci.*, 35, 690–713.

Schwiesow, R. L., 1981: Horizontal velocity structure in waterspouts. *J. Appl. Meteor.*, 20, 349–360.

Schwiesow, R. L., R. E. Cupp, P. C. Sinclair, and R. F. Abbey, 1981: Waterspout velocity measurements by airborne Doppler lidar. *J. Appl. Meteor.*, 20, 341–348.

Scorer, R., 1972: *Clouds of the World.* David and Charles, Newton Abbot, Devon, Great Britain, 176 pp.

Seimon, A., 1993: Anomalous cloud-to-ground lightning in an F5-tornado-producing supercell thunderstorm on 28 August 1990. *Bull. Amer. Meteor. Soc.*, 74, 189–203.

Severe Storms Research Group of Saint Louis University, 1970: F. C. Bates' conceptual thoughts on severe thunderstorms. *Bull. Amer. Meteor. Soc.*, 51, 481–488.

Simpson, J., B. R. Morton, M. C. McCumber, and R. S. Penc, 1986: Observations and mechanisms of GATE waterspouts. *J. Atmos. Sci.*, 43, 753–782.

Simpson, J., G. Roff, B. R. Morton, K. Labas, G. Dietachmeyer, M. McCumber, and R. Penc, 1991: The Great Salt Lake waterspout. *Mon. Wea. Rev.*, 119, 2471–2770.

Sinclair, P. C., 1969: General characteristics of dust devils. *J. Appl. Meteor.*, 8, 32–45.

Smith, J. S., 1965: The hurricane-tornado. *Mon. Wea. Rev.*, 93, 453–459.

Smith, M., 1974: Visual observations of a tornadic thunderstorm. *Weatherwise*, 27, 256–258.

Smith, R. K., and L. M. Leslie, 1979: A numerical study of tornadogenesis in a rotating thunderstorm. *Quart. J. Roy. Meteor. Soc.*, 105, 107–127.

Smith, R. L., and D. W. Holmes, 1961: Use of Doppler radar in meteorological observations. *Mon. Wea. Rev.*, 89, 1–7.

Snow, J. T., 1982: A review of recent advances in tornado vortex dynamics. *Rev. Geophys. Space Phys.*, 20, 953–964.

Snow, J. T., 1984: The tornado. *Sci. Amer.*, 250, 86–97.

Snow, J. T., 1984: On the formation of particle sheaths in columnar vortices. *J. Atmos. Sci.*, 41, 2477–2491.

Snow, J. T., 1987: Atmospheric columnar vortices. *Rev. Geophys.*, 25, 371–385.

Snow, J. T., and R. L. Pauley, 1984: On the thermodynamic method for estimating maximum tornado windspeeds. *J. Clim. Appl. Meteor.*, 23, 1465–1468.

Snow, J. T., A. L. Wyatt, A. K. McCarthy, and E. Bishop, 1995: Fallout of debris from tornadic thunderstorms: A historical perspective and two examples from VORTEX. *Bull. Amer. Meteor. Soc.*, **76**, 1777–1790.

Squires, P., and J. S. Turner, 1962: An entraining jet model for cumulonimbus updraughts. *Tellus*, **14**, 422–434.

Staley, D. O., and R. L. Gall, 1979: Barotropic instability in a tornado vortex. *J. Atmos. Sci.*, **36**, 973–981.

Sterling, B., 1994: *Heavy Weather*. Bantam Books, New York, 310 pp.

Stout, G. E., and F. A. Huff, 1953: Radar records Illinois tornadogenesis. *Bull. Amer. Meteor. Soc.*, **34**, 281–284.

Taylor, W. L., 1973: Electromagnetic radiation from severe storms in Oklahoma during April 29–30, 1970. *J. Geophys. Res.*, **78**, 8761–8777.

Tepper, M., and W. E. Eggert, 1956: Tornado proximity traces. *Bull. Amer. Meteor. Soc.*, **37**, 152–159.

Thorarinsson, S., and B. Vonnegut, 1964: Whirlwinds produced by the eruption of Surtsey volcano. *Bull. Amer. Meteor. Soc.*, **45**, 440–444.

Turner, J. S., 1962: The "starting plume" in neutral surroundings. *J. Fluid Mech.*, **13**, 356–368.

Turner, J. S., and D. K. Lilly, 1963: The carbonated water tornado vortex. *J. Atmos. Sci.*, **20**, 468–471.

van Tassel, E. L., 1955: The North Platte Valley tornado outbreak of June 27, 1955. *Mon. Wea. Rev.*, **83**, 255–264.

Vonnegut, B., 1960: Electrical theory of tornadoes. *J. Geophys. Res.*, **65**, 203–212.

Vonnegut, B., 1975: Chicken plucking as measure of tornado wind speed. *Weatherwise*, **28**, 217.

Wakimoto, R. M., 1983: The West Bend, Wisconsin storm of 4 April 1981: A problem in operational meteorology. *J. Clim. Appl. Meteor.*, **22**, 181–189.

Wakimoto, R. M., and N. T. Atkins, 1996: Observations on the origins of rotation: The Newcastle tornado during VORTEX-94. *Mon. Wea. Rev.*, **124**, 384–407.

Wakimoto, R. M., and B. E. Martner, 1992: Observations of a Colorado tornado. Part II: Combined photogrammetric and Doppler radar analysis. *Mon. Wea. Rev.*, **120**, 522–543.

Wakimoto, R. M., and J. W. Wilson, 1989: Non-supercell tornadoes. *Mon. Wea. Rev.*, **117**, 1113–1140.

Ward, N. B., 1972: The exploration of certain features of tornado dynamics using a laboratory model. *J. Atmos. Sci.*, **29**, 1194–1204.

Wegener, A., 1928: Beitrage zur Mechanik der Tromben und Tornados. *Meteorol. Z.*, **45**, 201–214.

Weisman, M. L., 1993: The genesis of severe, long-lived bow echoes. *J. Atmos. Sci.*, **50**, 645–670.

Weisman, M. L., 1996: On the use of vertical wind shear versus helicity in interpreting supercell dynamics. In *Preprints, 18th Conference on Severe Local Storms*, American Meteorological Society, Boston, 200–204.

Weisman, M. L., and J. B. Klemp, 1982: The dependence of numerically simulated convective storms on vertical wind shear and buoyancy. *Mon. Wea. Rev.*, **110**, 504–520.

Weisman, M. L., and J. B. Klemp, 1984: The structure and classification of numerically simulated convective storms in directionally varying wind shears. *Mon. Wea. Rev.*, **112**, 2479–2498.

Weisman, M. L., and J. B. Klemp, 1986: Characteristics of isolated convective storms. *Mesoscale Meteorology and Forecasting*, American Meteorological Society, Boston, 331–358.

Weller, N., and P. J. Waite, 1969: The Weller method: Tornado detection by television. *Preprints, 6th Conference on Severe Local Storms*, American Meteorological Society, Boston, 169–171.

Wicker, L. J., 1996: The role of near surface wind shear on low-level mesocyclone generation and tornadoes. In *Preprints, 18th Conference on Severe Local Storms*, American Meteorological Society, Boston, 115–119.

Wicker, L. J., and R. B. Wilhelmson, 1995: Simulation and analysis of tornado development and decay within a three-dimensional supercell thunderstorm. *J. Atmos. Sci.*, **52**, 2675–2703.

Wilhelmson, R. B., 1974: The life cycle of a thunderstorm in three dimensions. *J. Atmos. Sci.*, **31**, 1629–1651.

Wilhelmson, R. B., and J. B. Klemp, 1978: A three-dimensional numerical simulation of splitting that leads to long-lived storms. *J. Atmos. Sci.*, **35**, 1037–1063.

Wilhelmson, R. B., and J. B. Klemp, 1981: A three-dimensional numerical simulation of splitting severe storms on 3 April 1964. *J. Atmos. Sci.*, **38**, 1581–1600.

Wilson, J. W., 1986: Tornadogenesis by non-precipitation induced wind shear lines. *Mon. Wea. Rev.*, **114**, 270–284.

World Meteorological Organization, 1970: Unusual weather in 1969. Part I: Europe and Asia. *WMO Bull.*, **19**, 109–115.

Wurman, J., 1994: Vector winds from a single-transmitter bistatic dual-Doppler radar network. *Bull. Amer. Meteor. Soc.*, **75**, 983–994.

Zrnic, D. S., and R. J. Doviak, 1975: Velocity spectra of vortices scanned with a pulse-Doppler radar. *J. Appl. Meteor.*, **14**, 1531–1539.

Index